Advances in Intelligent Systems and Computing

Volume 270

Series editor

Janusz Kacprzyk, Polish Academy of Sciences, Warsaw, Poland
e-mail: kacprzyk@ibspan.waw.pl

For further volumes:
http://www.springer.com/series/11156

About this Series

The series "Advances in Intelligent Systems and Computing" contains publications on theory, applications, and design methods of Intelligent Systems and Intelligent Computing. Virtually all disciplines such as engineering, natural sciences, computer and information science, ICT, economics, business, e-commerce, environment, healthcare, life science are covered. The list of topics spans all the areas of modern intelligent systems and computing.

The publications within "Advances in Intelligent Systems and Computing" are primarily textbooks and proceedings of important conferences, symposia and congresses. They cover significant recent developments in the field, both of a foundational and applicable character. An important characteristic feature of the series is the short publication time and world-wide distribution. This permits a rapid and broad dissemination of research results.

Young Im Cho · Donghan Kim
Eric T. Matson
Editors

Soft Computing in Artificial Intelligence

Springer

Editors
Young Im Cho
University of Suwon
Gyeonggi-do
Korea

Eric T. Matson
Purdue University
West Lafayette, IN
USA

Donghan Kim
Kyunghee University
Gyeonggi-do
Korea

ISSN 2194-5357 ISSN 2194-5365 (electronic)
ISBN 978-3-319-05514-5 ISBN 978-3-319-05515-2 (eBook)
DOI 10.1007/978-3-319-05515-2
Springer Cham Heidelberg New York Dordrecht London

Library of Congress Control Number: 2014933213

Printed on acid-free paper

Springer is part of Springer Science+Business Media (www.springer.com)

Preface

This book explores the concept of artificial intelligence based on knowledge-based algorithms. Given the current hardware and software technologies and artificial intelligence theories, we can think of how efficient to provide a solution, how best to implement a model and how successful to achieve it. This edition provides readers with the most recent progress and novel solutions in artificial intelligence. This book aims at presenting the research results and solutions of applications in relevance with artificial intelligence technologies. We propose to researchers and practitioners some methods to advance the intelligent systems and apply artificial intelligence to specific or general purpose. This book consists of 13 contributions that feature fuzzy (r, s)-minimal pre- and β-open sets, handling big coocurrence matrices, Xie-Beni-type fuzzy cluster validation, fuzzy c-regression models, combination of genetic algorithm and ant colony optimization, building expert system, fuzzy logic and neural network, individual role adaptation for team sports, application of polynomial neural networks, recursive neuro-fuzzy algorithm for water management, application of interactive genetic algorithm, and Artificial Neural Network (ANN) model. This edition is published in original, peer reviewed contributions covering from initial design to final prototypes and verifications.

To help readers understand articles, we describe the short introduction of each article as follows;
1. "Fuzzy (r, s)-minimal preopen sets and fuzzy (r, s)-M precontinuous mappings on (r, s)-fuzzy minimal spaces": This article introduces the concept of fuzzy (r, s)-minimal preopen set on an (r, s)-fuzzy minimal space. The authors also introduce the concept of fuzzy (r, s)-M precontinuous mapping which is a generalization of fuzzy (r, s)-M continuous mapping, and investigate characterization of fuzzy (r, s)-M precontinuity.
2. "On fuzzy (r, s)-minimal β-open sets on (r, s)-fuzzy minimal spaces": This paper introduces the concept of fuzzy (r, s)-minimal β-open set on an (r, s)-fuzzy minimal space and basic some properties. The authors also introduce the concept of fuzzy (r, s)-M β-continuous mapping which is a generalization of fuzzy (r, s)-M semicontinuous mapping and fuzzy (r, s)-M precontinuous mapping, and investigate characterization for such mappings.
3. "Handling Very Large Cooccurrence Matrices in Fuzzy Co-clustering by Sampling

Approaches": This article presents how to apply the sampling approaches to fuzzy co-clustering tasks for handling cooccurrence matrices composed of many objects. In some modified approaches, item memberships are utilized in conjunction with other objects for inheriting the property of other sample sets.
4. "Xie-Beni-type Fuzzy Cluster Validation in Fuzzy Co-clustering of Documents and Keywords": This paper describes the applicability of a Xie-Beni-type co-cluster validity index to Fuzzy Co-clustering of Documents and Keywords (Fuzzy CoDoK). In Xie-Beni-type indices, intra-cluster deviations and inter-cluster distance are able to measure cluster compactness and separateness.
5. "Fuzzy c-Regression Models Based on Optimal Scaling of Categorical Observation with Tolerance": In data mining applications, the authors often deal with databases consisting of mixed measurement levels. In order to handle uncertain categorical observations, this article proposes to modify Fuzzy c-Regression Models (FCRM) as an FCM-type switching regression model.
6. "Combination of Genetic Algorithm and Ant Colony Optimization for QoS Multicast Routing": This paper presents a new method GAACO to solve this minimum cost multicast routing problem. In this method, genetic algorithm (GA) and ant colony optimization (ACO) are combined to improve the computing performance. The simulation results show that the proposed GAACO algorithm has superior performance when compared to other existing algorithms.
7. "Building a Knowledge Base for Expert System in Information Security": This paper asserts that expert systems approach can be beneficial in achieving this automation. Although information security is a very broad field, encompassing many complex concepts, the authors are trying to develop a methodology of formalizing of IS knowledge to build a knowledge base for expert system that can serve as IS audit expert.
8. "Fuzzy logic and neural network based induction control in a diesel engine": In this paper, the combination of fuzzy and neural network approach will help to avoid the high emissions and unexpected uncomfortable driving mode for the drivers. This new control approach, which uses the hybrid approach of fuzzy logic and neural network, is very easy to tune, simplify the development time, improve the control precision of the air system and reduce cost and time of calibration.
9. "Evaluation and Analysis of Relationship and Individual Role Adaptation for Team Sports using SOM": This paper reports the concept of the algorithm and result of physical or psychological data analysis using self-organizing maps and principal component analysis, respectively. This research aims to develop a clustering method that considers both physical and psychological features in judging an individual's role and adaptation in the game.
10. "Design of Radial Basis Function Classifier based on Polynomial Neural Networks": In order to improve the generalization ability of radial basis function networks, this paper shows a way to apply the polynomial neural networks as the virtual input variables of radial basis function networks.
11. "Recursive Neuro-Fuzzy Algorithm for Flow Prediction and Pump On-Off Minimization": This article proposes a new integrated solution for predictions and optimal pump control by learning algorithms. Flow prediction has usually been studied for daily or monthly estimation, which is insufficient for real-time control of a water treatment

plant (hereafter WTP). The proposed on/off minimization algorithm is expected to prevent those particles from leaking and to securing public health.

12. “Logotype Design Support System Based on Sketches”: This paper proposes a support system for interactive logotypes design based on user’s sketches, which applies interactive genetic algorithm. It also explains two types of experiments for the confirmation of proposed system validity: whether the system supports logotype design, and whether sketch inputs are useful. Experimental results show the usefulness of the proposed system.

13. “An Artificial Neural Network Based on the Architecture of the Cerebellum for Behavior Learning”: This article focuses on the human brain architecture that gives it the ability to self-organize and generalize sensory information. The authors propose an Artificial Neural Network (ANN) model based on that architecture. The paper describes a cerebellum-based ANN model (C-ANN) and verifies its capacity to learn from the phototoxic behavior acquisition of a simple two-wheeled robot.

We would appreciate it if readers could get useful information from the articles and contribute to creating innovative and novel concept or theory. Thank you.

Editors
Young Im Cho
Donghan Kim
Eric T. Matson

Contents

Fuzzy (r, s)-Minimal Preopen Sets and Fuzzy (r, s)-M Precontinuous Mappings on (r, s)-Fuzzy Minimal Spaces

Young Key Kim[1] and Won Keun Min[2,⋆]

[1] Department of Mathematics, MyongJi University, Youngin 449-728, Korea
ykkim@mju.ac.kr

[2] Department of Mathematics, Kangwon National University, Chuncheon, 200-701, Korea
wkmin@kangwon.ac.kr

Abstract. We introduce the concept of fuzzy (r, s)-minimal preopen set on an (r, s)-fuzzy minimal space. We also introduce the concept of fuzzy (r, s)-M precontinuous mapping which is a generalization of fuzzy (r, s)-M continuous mapping, and investigate characterization of fuzzy (r, s)-M precontinuity.

Keywords: (r, s)-fuzzy minimal structures, fuzzy (r, s)-minimal open, fuzzy (r, s)-minimal preopen, fuzzy (r, s)-M continuous, fuzzy (r, s)-M precontinuous.

1 Introduction

The concept of fuzzy set was introduced by Zadeh [18]. Chang introduced the concept of fuzzy topology in terms of fuzzy sets defined by Zadeh. Kubiak [11] and Šostak [14] introduced the fundamental concept of a fuzzy topological structure, as an extension of both crisp topology and fuzzy topology [4]. In [16,17], Šostak gave some rules and showed how such an extension can be realized. Chattopadhyay et al., [5] have redefined the same concept under the name gradation of openness. A general approach to the study of topological type structures on fuzzy power sets was developed in [8-12].

As a generalization of fuzzy sets, the concept of intuitionistic fuzzy sets was introduced by Atanassov [3]. By using intuitionistic fuzzy sets, Çoker and his colleagues [6,7] introduced intuitionistic fuzzy topological spaces. In [5], Çoker and Demirci introduced intuitionistic fuzzy topological spaces in Šostak's sense as a generalization of smooth fuzzy topological spaces and intuitionistic fuzzy topological spaces. Recently, Samanta and Mondal [14], introduced the notion of intuitionistic gradation of openness of fuzzy sets, where to each fuzzy subsets there is a definite grade of openness and there is a grade of non-openness. Thus, the concept of intuitionistic gradation of openness is a generalization of the concept of gradation of openness and the topology of intuitionistic fuzzy sets. In [13], we introduced the concept of (r, s)-fuzzy minimal space which is an extension of the intuitionistic fuzzy topological space in Šostak's sense. We also introduced and studied the concept of fuzzy (r, s)-M continuity. In this paper, we introduce

⋆ Corresponding author. *2000 Mathematics Subject Classification*:54C08

Y.I. Cho et al. (eds.), *Soft Computing in Artificial Intelligence*,
Advances in Intelligent Systems and Computing 270,
DOI: 10.1007/978-3-319-05515-2_1,

the concepts of fuzzy (r,s)-minimal preopen sets, fuzzy (r,s)-minimal preinterior and fuzzy (r,s)-minimal preclosure operators. We also introduce the fuzzy (r,s)-M precontinuous mapping. They are the generalized concepts of fuzzy (r,s)-minimal open sets and fuzzy (r,s)-M continuous mappings. First, we study some related properties for the fuzzy (r,s)-minimal preinterior and fuzzy (r,s)-minimal preclosure operators. In particular, we investigate some characterizations for the fuzzy (r,s)-M precontinuous mapping in terms of fuzzy (r,s)-minimal interior and fuzzy (r,s)-minimal closure operators.

2 Preliminaries

Let I be the unit interval $[0,1]$ of the real line. A member μ of I^X is called a fuzzy set of X. For any $\mu \in I^X$, μ^c denotes the complement $1-\mu$. By $\underline{0}$ and $\underline{1}$ we denote constant maps on X with value 0 and 1, respectively. All other notations are standard notations of fuzzy set theory.

Let X be a nonempty set. An *intuitionistic fuzzy set* A is an ordered pair

$$A = (\mu_A, \gamma_A)$$

where the functions $\mu_A : X \to I$ and $\gamma_A : X \to I$ denote the degree of membership and the degree of nonmembership, respectively, and $\mu_A(x)+\gamma_A(x) \le 1$ for all $x \in X$.

Obviously every fuzzy set μ on X is an intuitionistic fuzzy set of the form $(\mu, \underline{1}-\mu)$.

Throughout this paper, let X be a nonempty set, $I = [0,1]$ and $I_0 = (0,1]$ and $I_1 = [0,1)$. For $\alpha \in I$, $\underline{\alpha}(x) = \alpha$ for all $x \in X$.

Definition 1 ([14]). *An intuitionistic gradation of openness (IGO, for short) on X is an ordered pair (τ,τ^*) of functions from I^X to I such that*

(IGO1) $\tau(\lambda)+\tau^*(\lambda) \le 1, \forall \lambda \in I^X$,

(IGO2) $\tau(\underline{0}) = \tau(\underline{1}) = 1, \tau^*(\underline{0}) = \tau^*(\underline{1}) = 0$,

(IGO3) $\tau(\lambda_1 \wedge \lambda_2) \ge \tau(\lambda_1) \wedge \tau(\lambda_2)$ and $\tau^*(\lambda_1 \wedge \lambda_2) \le \tau^*(\lambda_1) \vee \tau^*(\lambda_2)$, for each $\lambda_1, \lambda_2 \in I^X$,

(IGO4) $\tau(\bigvee_{i\in\Delta} \lambda_i) \ge \bigwedge_{i\in\Delta} \tau(\lambda_i)$ and $\tau^*(\bigvee_{i\in\Delta} \lambda_i) \le \bigvee_{i\in\Delta} \tau^*(\lambda_i)$, for each $\lambda_i \in I^X, i \in \Delta$.

The triplet (X,τ,τ^*) is called an intuitionistic fuzzy topological space (IFTS, for short). τ and τ^* may be interpreted as gradation of openness and gradation of nonopenness, respectively.

Definition 2 ([13]). *Let X be a nonempty set, $r \in I_0$, $s \in I_1$ and $r+s \le 1$. The pair $(\mathcal{M},\mathcal{M}^*)$ of maps $\mathcal{M},\mathcal{M}^* : I^X \to I$ on X is said to have an (r,s)-fuzzy minimal structure if the family*

$$\mathcal{M}_{r,s} = \{\lambda \in I^X \mid \mathcal{M}(\lambda) \ge r \text{ and } \mathcal{M}^*(\lambda) \le s\}$$

contains $\underline{0}$ and $\underline{1}$. Then $(X,\mathcal{M},\mathcal{M}^*)$ is called an (r,s)-fuzzy minimal space (simply, (r,s)-FMS).

Then every member of $\mathcal{M}_{(r,s)}$ is called a fuzzy (r,s)-minimal open set. A fuzzy set λ is called a fuzzy (r,s)-minimal closed set if the complement of λ is a fuzzy (r,s)-minimal open set.

Definition 3 ([13]). *Let $(X, \mathcal{M}, \mathcal{M}^*)$ be an (r, s)-FMS. The (r, s)-fuzzy minimal closure and (r, s)-fuzzy minimal interior of λ, denoted by $\mathcal{C}_m(\lambda, r, s)$ and $\mathcal{I}_m(\lambda, r, s)$, respectivity, are defined as*

$$\mathcal{C}_m(\lambda, r, s) = \bigwedge\{\mu \in I^X \mid \underline{1} - \mu \in \mathcal{M}_{r,s} \text{ and } \lambda \leq \mu\},$$

$$\mathcal{I}_m(\lambda, r, s) = \bigvee\{\mu \in I^X \mid \mu \in \mathcal{M}_{r,s} \text{ and } \lambda \geq \mu\}.$$

Theorem 1 ([13]). *Let $(X, \mathcal{M}, \mathcal{M}^*)$ be an (r, s)-FMS and $\lambda, \mu \in I^X$. Then*
(1) $\mathcal{I}_m(\lambda, r, s) \leq \lambda$ and if $\lambda \in \mathcal{M}_{r,s}$ then $\mathcal{I}_m(\lambda, r, s) = \lambda$.
(2) $\mathcal{C}_m(\lambda, r, s) \geq \lambda$ and if $\underline{1} - \lambda \in \mathcal{M}_{r,s}$ then $\mathcal{C}_m(\lambda, r, s) = \lambda$.
(3) If $\lambda \leq \mu$ then $\mathcal{I}_m(\lambda, r, s) \leq \mathcal{I}_m(\mu, r, s)$ and $\mathcal{C}_m(\lambda, r, s) \leq \mathcal{C}_m(\mu, r, s)$.
(4) $\mathcal{I}_m(\lambda \wedge \mu, r, s) = \mathcal{I}_m(\lambda, r, s) \wedge \mathcal{I}_m(\mu, r, s)$ and $\mathcal{C}_m(\lambda \vee \mu, r, s) = \mathcal{C}_m(\lambda, r, s) \vee \mathcal{C}_m(\mu, r, s)$.
(5) $\mathcal{I}_m(\mathcal{I}_m(\lambda, r, s), r, s) = \mathcal{I}_m(\lambda, r, s)$ and $\mathcal{C}_m(\mathcal{C}_m(\lambda, r, s), r, s) = \mathcal{C}_m(\lambda, r, s)$.
(6) $\underline{1} - \mathcal{C}_m(\lambda, r, s) = \mathcal{I}_m(\underline{1} - \lambda, r, s)$ and $\underline{1} - \mathcal{I}_m(\lambda, r, s) = \mathcal{C}_m(\underline{1} - \lambda, r, s)$

Definition 4 ([13]). *Let $(X, \mathcal{M}, \mathcal{M}^*)$ and $(Y, \mathcal{N}, \mathcal{N}^*)$ be two (r, s)-FMS's. Then $f : X \to Y$ is said to be (r, s)-fuzzy M-continuous if for every $\mu \in \mathcal{N}_{r,s}$, $f^{-1}(\mu) \in \mathcal{M}_{r,s}$.*

Theorem 2 ([13]). *Let $f : (X, \mathcal{M}, \mathcal{M}^*) \to (Y, \mathcal{N}, \mathcal{N}^*)$ be a function,*
(1) f is (r, s)-fuzzy M-continuous.
(2) $\underline{1} - f^{-1}(\mu) \in \mathcal{M}_{r,s}$, for each $\underline{1} - \mu \in \mathcal{N}_{r,s}$ and $\mu \in I^Y$.
(3) $f(\mathcal{C}_m(\lambda, r, s)) \leq \mathcal{C}_m(f(\lambda), r, s)$, for $\lambda \in I^X$.
(4) $\mathcal{C}_m(f^{-1}(\mu), r, s)) \leq f^{-1}(\mathcal{C}_m(\mu, r, s))$, for $\mu \in I^Y$.
(5) $f^{-1}(\mathcal{I}_m(\mu, r, s)) \leq \mathcal{I}_m(f^{-1}(\mu), r, s)$, for $\mu \in I^Y$.
Then, $(1) \Leftrightarrow (2) \Rightarrow (3) \Leftrightarrow (4) \Leftrightarrow (5)$.

3 Fuzzy (r, s)-Minimal Preopen Sets and Fuzzy (r, s)-M Precontinuity

Definition 5. *Let $(X, \mathcal{M}, \mathcal{M}^*)$ be an (r, s)-FMS and $\lambda \in I^X$. Then a fuzzy set λ is called a fuzzy (r, s)-minimal preopen set in X if*

$$\lambda \leq mI(mC(\lambda, r, s), r, s).$$

A fuzzy set λ is called a fuzzy (r, s)-minimal preclosed set if the complement of λ is fuzzy (r, s)-minimal preopen.

Every fuzzy (r, s)-minimal open set is fuzzy (r, s)-minimal preopen but the converse may not be true in general.

Example 1. Let $X = I$. For $0 < n < 1$, consider the following fuzzy sets

$$\mu_n(x) = \begin{cases} \frac{x}{n} & \text{if } \ 0 < x \leq n \\ \frac{1-x}{1-n} & \text{if } \ n < x \leq 1; \end{cases}$$

$$\lambda(x) = \begin{cases} \frac{x}{2} + \frac{3}{4} & \text{if } 0 \leq x \leq \frac{1}{2} \\ 1 & \text{if } \frac{1}{2} < x \leq 1. \end{cases}$$

Define $\mathcal{M}, \mathcal{M}^* : I^X \to I$ on X by

$$\mathcal{M}(\mu) = \begin{cases} n & \text{if } \mu = \mu_n \\ \max(\{\alpha, 1-\alpha\}) & \text{if } \mu = \underline{\alpha} \\ 0 & \text{otherwise,} \end{cases}$$

$$\mathcal{M}^*(\mu) = \begin{cases} 1-n & \text{if } \mu = \mu_n \\ \min(\{1-\alpha, \alpha\}) & \text{if } \mu = \underline{\alpha} \\ 1 & \text{otherwise.} \end{cases}$$

Consider a $(\frac{3}{4}, \frac{1}{4})$-fuzzy minimal structure $(\mathcal{M}_{\frac{3}{4}}, \mathcal{M}^*_{\frac{1}{4}})$ as follows

$$\mathcal{M}_{\frac{3}{4},\frac{1}{4}} = \{\mu \in I^X \mid \mathcal{M}(\mu) \geq \frac{3}{4} \text{ and } \mathcal{M}^*(\mu) \leq \frac{1}{4}\}.$$

Then

$$\mathcal{I}_m(\lambda, \frac{3}{4}, \frac{1}{4}) = \begin{cases} 2x & \text{if } 0 \leq x \leq \frac{1}{2} \\ 1 & \text{if } \frac{1}{2} < x < 1 \\ 0 & \text{if } x = 1, \end{cases}$$

and $\mathcal{C}_m(\lambda, \frac{3}{4}, \frac{1}{4}) = \underline{1}$.

So λ is fuzzy (r, s)-minimal preopen but it is not fuzzy (r, s)-minimal open.

Remark 1. Let $(X, \mathcal{M}, \mathcal{M}^*)$ be an (r, s)-FMS. Then

(1) a fuzzy set λ is fuzzy (r, s)-minimal preclosed if and only if $mC(mI(\lambda, r, s), r, s) \leq \lambda$.

(2) $\underline{0}$ and $\underline{1}$ are both fuzzy (r, s)-minimal preclosed.

(3) If $(X, \mathcal{M}, \mathcal{M}^*)$ is an (r, s)-IFMS, then every fuzzy (r, s)-minimal preopen set is exactly a fuzzy (r, s)-preopen set [2].

Theorem 3. *Let $(X, \mathcal{M}, \mathcal{M}^*)$ be an (r, s)-FMS. Any union of fuzzy (r, s)-minimal preopen sets is fuzzy (r, s)-minimal preopen.*

Proof. Let λ_i be a fuzzy (r, s)-minimal preopen set for $i \in J$. Then from $\lambda_i \leq \vee\lambda_i$,

$$\lambda_i \leq mI(mC(\lambda_i, r, s), r, s) \leq mI(mC(\vee\lambda_i, r, s), r, s).$$

This implies $\vee\lambda_i \leq mI(mC(\vee\lambda_i, r, s), r, s)$ and so $\vee\lambda_i$ is fuzzy (r, s)-minimal preopen. □

In the next example, we can show that the intersection of two fuzzy (r, s)-minimal preopen sets is not fuzzy (r, s)-minimal preopen in general.

Example 2. As in Examle 1, let us consider a fuzzy $(\frac{3}{4}, \frac{1}{4})$-minimal structure $(\mathcal{M}_{\frac{3}{4}}, \mathcal{M}^*_{\frac{1}{4}})$. Consider two fuzzy sets λ and γ defined as the following:

$$\lambda(x) = \begin{cases} -2x + \frac{1}{2} & \text{if } 0 \leq x \leq \frac{1}{4} \\ 1 & \text{if } \frac{1}{4} < x < \frac{3}{4} \\ 2x - \frac{3}{2} & \text{if } \frac{3}{4} \leq x \leq 1. \end{cases}$$

$$\gamma(x) = \begin{cases} -2x+1 & \text{if } 0 \le x \le \frac{1}{4} \\ 0 & \text{if } \frac{1}{4} < x < \frac{3}{4} \\ 2x-1 & \text{if } \frac{3}{4} \le x \le 1. \end{cases}$$

Then obviously λ and γ are fuzzy $(\frac{3}{4},\frac{1}{4})$–minimal preopen. But since $\lambda \wedge \gamma \not\leq mI(mC(\lambda \wedge \gamma, \frac{3}{4}, \frac{1}{4}), \frac{3}{4}, \frac{1}{4}) = \underline{0}$, $\lambda \wedge \gamma$ is not fuzzy $(\frac{3}{4},\frac{1}{4})$-minimal preopen.

Definition 6. *Let $(X, \mathcal{M}, \mathcal{M}^*)$ be an (r,s)-FMS. For $\lambda \in I^X$, $mpC(\lambda, r, s)$ and $mpI(\lambda, r, s)$ are defined as the following:*

$$mpC(\lambda, r, s) = \wedge\{\gamma \in I^X : \lambda \le \gamma,\ \ \gamma \textit{ is fuzzy } (r,s)\textit{-minimal preclosed}\}$$

$$mpI(\lambda, r, s) = \vee\{\mu \in I^X : \mu \le \lambda,\ \ \mu \textit{ is fuzzy } (r,s)\textit{-minimal preopen}\}.$$

Theorem 4. *Let $(X, \mathcal{M}, \mathcal{M}^*)$ be an (r,s)-FMS and $\lambda \in I^X$. Then*
(1) $mpI(\lambda, r, s) \le \lambda$.
(2) If $\lambda \le \mu$, then $mpI(\lambda, r, s) \le mpI(\mu, r, s)$.
(3) λ is (r,s)-minimal preopen iff $mpI(\lambda, r, s) = \lambda$.
(4) $mpI(pmI(\lambda, r, s), r, s) = mpI(\lambda, r, s)$.
(5) $mpC(\underline{1} - \lambda, r, s) = \underline{1} - mpI(\lambda, r, s)$ and $mpI(\underline{1} - \lambda, r, s) = \underline{1} - mpC(\lambda, r, s)$.

Proof. (1), (2), (3) and (4) are obviously obtained from Theorem 3.
(5) For $\lambda \in I^X$,

$$\begin{aligned} \underline{1} - mpI(\lambda, r, s) &= \underline{1} - \vee\{\mu \in I^X : \mu \le \lambda,\ \ \mu \text{ is fuzzy } (r,s)\text{-minimal preopen}\} \\ &= \wedge\{\underline{1} - \mu : \mu \le \lambda,\ \ \mu \text{ is fuzzy } (r,s)\text{-minimal preopen}\} \\ &= \wedge\{\underline{1} - \mu : \underline{1} - \lambda \le \underline{1} - \mu,\ \ \mu \text{ is fuzzy } (r,s)\text{-minimal preopen}\} \\ &= mpC(\underline{1} - \lambda, r, s). \end{aligned}$$

Similarly, we have $mpI(\underline{1} - \lambda, r, s) = \underline{1} - mpC(\lambda, r, s)$. □

Theorem 5. *Let $(X, \mathcal{M}, \mathcal{M}^*)$ be an (r,s)-FMS and $\lambda \in I^X$. Then*
(1) $\lambda \le mpC(\lambda, r, s)$.
(2) If $\lambda \le \mu$, then $mpC(\lambda, r, s) \le mpC(\mu, r, s)$.
(3) μ is (r,s)-minimal preclosed iff $mpC(\mu, r, s) = F$.
(4) $mpC(mpC(\lambda, r, s), r, s) = mpC(\lambda, r, s)$.

Proof. It is similar to the proof of Theorem 4. □

Definition 7. *Let $(X, \mathcal{M}, \mathcal{M}^*)$ and $(Y, \mathcal{N}, \mathcal{N}^*)$ be r-FMS's. Then a mapping $f : (X, \mathcal{M}, \mathcal{M}^*) \to (Y, \mathcal{N}, \mathcal{N}^*)$ is said to be fuzzy (r,s)-M precontinuous if for each fuzzy (r,s)-minimal open set $\lambda \in I^Y$, $f^{-1}(\lambda)$ is a fuzzy (r,s)-minimal preopen set.*

Every fuzzy (r,s)-M continuous mapping is fuzzy (r,s)-M precontinuous but the converse is not true in general.

Example 3. Let $X = I$. For $0 < n < 1$, consider the following fuzzy sets

$$\mu_n(x) = \begin{cases} \frac{x}{n} & \text{if } 0 < x \le n \\ \frac{1-x}{1-n} & \text{if } n < x \le 1; \end{cases}$$

$$\lambda(x) = \begin{cases} \frac{x}{2} + \frac{3}{4} & \text{if } 0 \le x \le \frac{1}{2} \\ 1 & \text{if } \frac{1}{2} < x \le 1. \end{cases}$$

Define $\mathcal{M}, \mathcal{M}^* : I^X \to I$ on X by

$$\mathcal{M}(\mu) = \begin{cases} n & \text{if } \mu = \mu_n \\ \max(\{\alpha, 1-\alpha\}) & \text{if } \mu = \underline{\alpha} \\ 0 & \text{otherwise,} \end{cases}$$

$$\mathcal{M}^*(\mu) = \begin{cases} 1-n & \text{if } \mu = \mu_n \\ \min(\{1-\alpha, \alpha\}) & \text{if } \mu = \underline{\alpha} \\ 1 & \text{otherwise.} \end{cases}$$

Define $\mathcal{N}, \mathcal{N}^* : I^X \to I$ on X by

$$\mathcal{N}(\mu) = \begin{cases} n & \text{if } \mu = \mu_n \\ \max(\{\alpha, 1-\alpha\}) & \text{if } \mu = \underline{\alpha} \\ \frac{7}{8} & \text{if } \mu = \lambda \\ 0 & \text{otherwise,} \end{cases}$$

$$\mathcal{N}^*(\mu) = \begin{cases} 1-n & \text{if } \mu = \mu_n \\ \min(\{1-\alpha, \alpha\}) & \text{if } \mu = \underline{\alpha} \\ 0 & \text{if } \mu = \lambda \\ 1 & \text{otherwise.} \end{cases}$$

Note that $\lambda \in \mathcal{N}_{\frac{3}{4},\frac{1}{4}}$, $\lambda \notin \mathcal{M}_{\frac{3}{4},\frac{1}{4}}$ and λ is fuzzy (r,s)-minimal preopen in $(X, \mathcal{M}, \mathcal{M}^*)$. Finally we can say the identity mapping $f : (X, \mathcal{M}, \mathcal{M}^*) \to (X, \mathcal{N}, \mathcal{N}^*)$ is fuzzy (r,s)-M precontinuous but not fuzzy (r,s)-M continuous.

Theorem 6. *Let $f : (X, \mathcal{M}, \mathcal{M}^*) \to (Y, \mathcal{N}, \mathcal{N}^*)$ be a mapping on (r,s)-FMS's $(X, \mathcal{M}, \mathcal{M}^*)$ and $(Y, \mathcal{N}, \mathcal{N}^*)$. Then the following statements are equivalent:*

(1) f is fuzzy (r,s)-M precontinuous.

(2) $f^{-1}(\gamma)$ is a fuzzy (r,s)-minimal preclosed set for each fuzzy (r,s)-minimal closed set $\gamma \in I^Y$.

(3) $f(mpC(\lambda, r, s)) \le mC(f(\lambda), r, s)$ for $\lambda \in I^X$.

(4) $mpC(f^{-1}(\mu), r, s) \le f^{-1}(mC(\mu, r, s))$ for $\mu \in I^Y$.

(5) $f^{-1}(mI(\mu, r, s)) \le mpI(f^{-1}(\mu), r, s)$ for $\mu \in I^Y$.

Proof. $(1) \Rightarrow (2)$ It is obvious.

$(2) \Rightarrow (3)$ For $\lambda \in I^X$,

$$\begin{aligned} f^{-1}(mC(f(\lambda), r, s)) &= f^{-1}(\wedge\{\gamma \in I^Y : f(\lambda) \le \gamma \text{ and } \gamma \text{ is fuzzy } (r,s)\text{-minimal closed}\}) \\ &= \wedge\{f^{-1}(\gamma) \in I^X : \lambda \le f^{-1}(\gamma) \text{ and} \\ &\quad f^{-1}(\gamma) \text{ is fuzzy } (r,s)\text{-minimal preclosed}\} \\ &\ge \wedge\{\mu \in I^X : \lambda \le \mu \text{ and } \mu \text{ is fuzzy } (r,s)\text{-minimal preclosed}\} \\ &= mpC(\lambda, r, s) \end{aligned}$$

Hence $f(mpC(\lambda, r, s)) \le mC(f(\lambda), r, s)$.

(3) ⇒ (4) For $\mu \in I^Y$,

$$f(mpC(f^{-1}(\mu)), r, s) \leq mC(f(f^{-1}(\mu)), r, s) \leq mC(\mu, r, s).$$

Thus this implies $mpC(f^{-1}(\mu), r, s) \leq f^{-1}(mC(\mu, r, s)$.

(4) ⇒ (5) For $\mu \in I^Y$, from Theorem 1,

$$\begin{aligned} f^{-1}(mI(\mu), r, s) &= f^{-1}(\underline{1} - mC(\underline{1} - \mu, r, s)) \\ &= \underline{1} - f^{-1}(mC(\underline{1} - \mu, r, s)) \\ &\leq \underline{1} - mpC(f^{-1}(\underline{1} - \mu), r, s) \\ &= mpI(f^{-1}(\mu), r, s). \end{aligned}$$

Hence $f^{-1}(mI(\mu, r, s)) \leq mpI(f^{-1}(\mu), r, s)$.

(5) ⇒ (1) Let λ be any fuzzy (r, s)-minimal open set. Then from (5), it follows $f^{-1}(\lambda) = f^{-1}(mI(\lambda, r, s)) \leq mpI(f^{-1}(\lambda), r, s)$ and $f^{-1}(\lambda) = mpI(f^{-1}(\lambda), r, s)$. This implies $f^{-1}(\lambda)$ is a fuzzy (r, s)-minimal preopen set. Hence f is fuzzy (r, s)-M precontinuous. □

Theorem 7. *Let $(X, \mathcal{M}, \mathcal{M}^*)$ be an (r, s)-FMS and $\lambda \in I^X$. Then*

(1) $mC(mI(\lambda, r, s), r, s) \leq mC(mI(mpC(\lambda, r, s), r, s), r, s) \leq mpC(\lambda, r, s)$.

(2) $mpI(\lambda, r, s) \leq mI(mC(mpI(\lambda, r, s), r, s), r, s) \leq mI(mC(\lambda, r, s), r, s)$.

Proof. (1) Since $mpC(\lambda, r, s)$ is fuzzy (r, s)-minimal preclosed, it is obtained from (1) of Remark 1.

(2) Obvious.

□

Theorem 8. *Let $f : (X, \mathcal{M}, \mathcal{M}^*) \to (Y, \mathcal{N}, \mathcal{N}^*)$ be a mapping on r-FMS's $(X, \mathcal{M}, \mathcal{M}^*)$ and $(Y, \mathcal{N}, \mathcal{N}^*)$. Then the following statements are equivalent:*

(1) f is fuzzy (r, s)-M precontinuous.

(2) $f^{-1}(\mu) \leq mI(mC(f^{-1}(\mu), r, s), r, s)$ for each fuzzy (r, s)-minimal open set μ in Y.

(3) $mC(mI(f^{-1}(\gamma), r, s), r, s) \leq f^{-1}(\gamma)$ for each fuzzy (r, s)-minimal closed set γ in Y.

(4) $f(mC(mI(\lambda, r, s), r, s)) \leq mC(f(\lambda), r, s)$ for $\lambda \in I^X$.

(5) $mC(mI(f^{-1}(\mu), r, s), r, s) \leq f^{-1}(mC(\mu, r, s))$ for $\mu \in I^Y$.

(6) $f^{-1}(mI(\mu, r, s)) \leq mI(mC(f^{-1}(\mu), r, s), r, s)$ for $\mu \in I^Y$.

Proof. (1) ⇔ (2) It is easily obtained from definition of fuzzy (r, s)-minimal preopen sets.

(1) ⇔ (3) Obvious.

(1) ⇒ (4) For $\lambda \in I^X$, we have

$$mC(mI(\lambda, r, s), r, s) \leq mpC(\lambda, r, s) \leq f^{-1}(f(mpC(\lambda, r, s))) \leq f^{-1}(mC(f(\lambda), r, s)).$$

So $f(mC(mI(\lambda, r, s), r, s)) \leq mC(f(\lambda), r, s)$.

(4) ⇒ (5) Obvious.

(5) $\Rightarrow$ (6) For $\mu \in I^Y$, from Theorem 1 and hypothesis,

$$\begin{aligned} f^{-1}(mI(\mu, r, s)) &= f^{-1}(\underline{1} - mC(\underline{1} - \mu, r, s)) \\ &= \underline{1} - (f^{-1}(mC(\underline{1} - \mu, r, s))) \\ &\leq \underline{1} - mC(mI(f^{-1}(\underline{1} - \mu), r, s), r, s) \\ &= mI(mC(f^{-1}(\mu), r, s), r, s). \end{aligned}$$

So we have (6).

(6) $\Rightarrow$ (1) For $\mu \in I^Y$, let μ be a fuzzy (r, s)-minimal open set. Then since $\mu = mI(\mu, r, s)$, by hypothesis, $f^{-1}(\mu) = f^{-1}(mI(\mu, r, s)) \leq mI(mC(f^{-1}(\mu), r, s), r, s)$ and so $f^{-1}(\mu)$ is fuzzy (r, s)-minimal preopen. Hence f is fuzzy (r, s)-M precontinuous. □

References

1. Abbas, S.E., Azab, M.A.: Some properties of R-fuzzy semiopen sets. J. Fuzzy Math. 13(2), 407–422 (2005)
2. Abbas, S.E., Krsteska, B.: Intuitionistic fuzzy strongly preopen (preclosed) mappings. Math. Morav 10, 47–53 (2006)
3. Atanassov, K.: Intuitionistic fuzzy sets. Fuzzy Sets and Systems 20(1), 87–96 (1986)
4. Chang, C.L.: Fuzzy topological spaces. J. Math. Anal. Appl. 24, 182–190 (1968)
5. Chattopadhyay, K.C., Hazra, R.N., Samanta, S.K.: Gradation of openess: fuzzy topology. Fuzzy Sets and Systems 49, 237–242 (1992)
6. Çoker: An introduction to intuitionistic fuzzy topological spaces. Fuzzy Sets and Systems 88, 81–89 (1997)
7. Çoker, D., Dimirci, M.: An introduction to intuitionistic fuzzy topological spaces in Šostak sense. Buseful 67, 67–76 (1996)
8. Höhle, U.: Upper semicontinuous fuzzy sets and applications. J. Math. Anall. Appl. 78, 659–673 (1980)
9. Höhle, U., Šostak, A.P.: A general theory of fuzzy topological spaces. Fuzzy Sets and Systems 73, 131–149 (1995)
10. Höhle, U., Šostak, A.P.: Axiomatic Foundations of Fixed-Basis fuzzy topology, ch. 3. The Handbooks of Fuzzy sets series, vol. 3. Kluwer Academic Publishers, Dordrecht (1999)
11. Kubiak, T.: On fuzzy topologies. Ph. D. Thesis, A. Mickiewicz, Poznan (1985)
12. Kubiak, T., Šostak, A.P.: Lower set-valued fuzzy topologies. Quaestions Math. 20(3), 423–429 (1997)
13. Min, W.K., Abbas, S.E.: (r, s)-fuzzy minimal structures and (r, s)-fuzzy minimal spaces. Journal Journal of Intelligent and Fuzzy Systems, doi:10.3233/IFS-120669 (accepted)
14. Samanta, S.K., Mondal, T.K.: On intuitionistic gradation of openness. Fuzzy Sets and Systems 131, 323–336 (2002)
15. Šostak, A.P.: On a fuzzy topological structure, Suppl. Rend. Circ. Matem. Palerms Ser. II 11, 89–103 (1985)
16. Šostak, A.P.: Two decades of fuzzy topology: basic ideas, notion and results. Russian Math. Surveys 44(6), 125–186 (1989)
17. Šostak, A.P.: Basic structures of fuzzy topology. J. Math. Sci. 78(6), 662–701 (1996)
18. Zadeh, L.A.: Fuzzy sets. Information and Control 8, 338–353 (1965)

On Fuzzy (r, s)-Minimal β-Open Sets on (r, s)-Fuzzy Minimal Spaces

Young Key Kim[1] and Won Keun Min[2,*]

[1] Department of Mathematics, MyongJi University, Youngin 449-728, Korea
ykkim@mju.ac.kr
[2] Department of Mathematics, Kangwon National University, Chuncheon, 200-701, Korea
wkmin@kangwon.ac.kr

Abstract. We introduce the concept of fuzzy (r,s)-minimal β-open set on an (r,s)-fuzzy minimal space and basic some properties. We also introduce the concept of fuzzy (r,s)-M β-continuous mapping which is a generalization of fuzzy (r,s)-M semicontinuous mapping and fuzzy (r,s)-M precontinuous mapping, and investigate characterization for such mappings.

Keywords: (r,s)-fuzzy minimal structures, fuzzy (r,s)-minimal open, fuzzy (r,s)-minimal β-open, fuzzy (r,s)-M β-continuous, fuzzy (r,s)-M^* β-open mapping, fuzzy (r,s)-M β-open mapping.

1 Introduction

The concept of fuzzy set was introduced by Zadeh [16]. Chang [2] introduced the concept of fuzzy topology in terms of fuzzy sets defined by Zadeh. Kubiak [9,10] and Šostak [15] also introduced the fundamental concept of a fuzzy topological structure. Chattopadhyay et al., [3] have redefined the same concept under the name gradation of openness. A general approach to the study of topological type structures on fuzzy power sets was developed in [6-10].

As a generalization of fuzzy sets, the concept of intuitionistic fuzzy sets was introduced by Atanassov [1]. By using intuitionistic fuzzy sets, Çoker and his colleagues [4,5] introduced intuitionistic fuzzy topological spaces. In [5], Çoker and Demirci introduced intuitionistic fuzzy topological spaces in Šostak's sense as a generalization of smooth fuzzy topological spaces and intuitionistic fuzzy topological spaces. Recently, Samanta and Mondal [14], introduced the notion of intuitionistic gradation of openness of fuzzy sets, where to each fuzzy subsets there is a definite grade of openness and there is a grade of non-openness. Thus, the concept of intuitionistic gradation of openness is a generalization of the concept of gradation of openness and the topology of intuitionistic fuzzy sets. In [11], Min and Abbas introduced the concept of (r,s)-fuzzy minimal space which is an extension of the intuitionistic fuzzy topological space in Šostak's sense. They also introduced and studied the concept of fuzzy (r,s)-M continuity. In [12,13], we introduced the concepts of fuzzy (r,s)-minimal semiopen set [13], fuzzy (r,s)-minimal preopen set [12], fuzzy (r,s)-M semicontinuous mapping [13]

* Corresponding author. *2000 Mathematics Subject Classification*:54C08

Y.I. Cho et al. (eds.), *Soft Computing in Artificial Intelligence*,
Advances in Intelligent Systems and Computing 270,
DOI: 10.1007/978-3-319-05515-2_2,

and fuzzy (r, s)-M precontinuous mapping [12]. We also studied some basic properties for such notions. In this paper, we are going to generalize the concepts of fuzzy (r, s)-minimal semiopen set and fuzzy (r, s)-minimal preopen set. So first, we introduce the concept of fuzzy (r, s)-minimal β-open set. Also we introduce the concepts of fuzzy (r, s)-M β-continuous mapping, fuzzy (r, s)-M^* β-open mapping and fuzzy (r, s)-M β-open mapping by using fuzzy (r, s)-minimal β-open sets. We study some basic properties for fuzzy (r, s)-minimal β-interior operators and fuzzy (r, s)-minimal β-closure operators. We also investigate characterizations for such mappings in terms of fuzzy (r, s)-minimal β-interior operators and fuzzy (r, s)-minimal β-closure operators.

2 Preliminaries

Let I be the unit interval $[0, 1]$ of the real line. A member μ of I^X is called a fuzzy set of X. For any $\mu \in I^X$, μ^c denotes the complement $1 - \mu$. By $\underline{0}$ and $\underline{1}$ we denote constant maps on X with value 0 and 1, respectively. All other notations are standard notations of fuzzy set theory.

Let X be a nonempty set. An *intuitionistic fuzzy set* A is an ordered pair

$$A = (\mu_A, \gamma_A)$$

where the functions $\mu_A : X \to I$ and $\gamma_A : X \to I$ denote the degree of membership and the degree of nonmembership, respectively, and $\mu_A(x) + \gamma_A(x) \leq 1$ for all $x \in X$.

Obviously every fuzzy set μ on X is an intuitionistic fuzzy set of the form $(\mu, \underline{1} - \mu)$.

Throughout this paper, let X be a nonempty set, $I = [0, 1]$ and $I_0 = (0, 1]$ and $I_1 = [0, 1)$. For $\alpha \in I$, $\underline{\alpha}(x) = \alpha$ for all $x \in X$.

Definition 1 ([14]). *An intuitionistic gradation of openness (IGO, for short) on X is an ordered pair (τ, τ^*) of functions from I^X to I such that*

(IGO1) $\tau(\lambda) + \tau^*(\lambda) \leq 1, \forall \lambda \in I^X$,

(IGO2) $\tau(\underline{0}) = \tau(\underline{1}) = 1, \tau^*(\underline{0}) = \tau^*(\underline{1}) = 0$,

(IGO3) $\tau(\lambda_1 \wedge \lambda_2) \geq \tau(\lambda_1) \wedge \tau(\lambda_2)$ *and* $\tau^*(\lambda_1 \wedge \lambda_2) \leq \tau^*(\lambda_1) \vee \tau^*(\lambda_2)$, *for each* $\lambda_1, \lambda_2 \in I^X$,

(IGO4) $\tau(\bigvee_{i\in\Delta} \lambda_i) \geq \bigwedge_{i\in\Delta} \tau(\lambda_i)$ *and* $\tau^*(\bigvee_{i\in\Delta} \lambda_i) \leq \bigvee_{i\in\Delta} \tau^*(\lambda_i)$, *for each* $\lambda_i \in I^X, i \in \Delta$.

The triplet (X, τ, τ^*) is called an intuitionistic fuzzy topological space (IFTS, for short). τ and τ^* may be interpreted as gradation of openness and gradation of nonopenness, respectively.

Definition 2 ([11]). *Let X be a nonempty set, $r \in I_0$, $s \in I_1$ and $r + s \leq 1$. The pair $(\mathcal{M}, \mathcal{M}^*)$ of maps $\mathcal{M}, \mathcal{M}^* : I^X \to I$ on X is said to have an (r, s)-fuzzy minimal structure if the family*

$$\mathcal{M}_{r,s} = \{\lambda \in I^X \mid \mathcal{M}(\lambda) \geq r \text{ and } \mathcal{M}^*(\lambda) \leq s\}$$

contains $\underline{0}$ and $\underline{1}$. Then $(X, \mathcal{M}, \mathcal{M}^*)$ is called an (r, s)-fuzzy minimal space (simply, (r, s)-FMS). Then every member of $\mathcal{M}_{(r,s)}$ is called a fuzzy (r, s)-minimal open set. A fuzzy set λ is called a fuzzy (r, s)-minimal closed set if the complement of λ is a fuzzy (r, s)-minimal open set.

Definition 3 ([11]). *Let $(X, \mathcal{M}, \mathcal{M}^*)$ be an (r, s)-FMS. The fuzzy (r, s)-minimal closure and fuzzy (r, s)-minimal interior of λ, denoted by $\mathcal{C}_m(\lambda, r, s)$ and $\mathcal{I}_m(\lambda, r, s)$, respectivity, are defined as*

$$\mathcal{C}_m(\lambda, r, s) = \wedge\{\mu \in I^X \mid \underline{1} - \mu \in \mathcal{M}_{r,s} \text{ and } \lambda \leq \mu\},$$

$$\mathcal{I}_m(\lambda, r, s) = \vee\{\mu \in I^X \mid \mu \in \mathcal{M}_{r,s} \text{ and } \lambda \geq \mu\}.$$

Theorem 1 ([11]). *Let $(X, \mathcal{M}, \mathcal{M}^*)$ be an (r, s)-FMS and $\lambda, \mu \in I^X$. Then*
(1) $\mathcal{I}_m(\lambda, r, s) \leq \lambda$ and if $\lambda \in \mathcal{M}_{r,s}$ then $\mathcal{I}_m(\lambda, r, s) = \lambda$.
(2) $\mathcal{C}_m(\lambda, r, s) \geq \lambda$ and if $\underline{1} - \lambda \in \mathcal{M}_{r,s}$ then $\mathcal{C}_m(\lambda, r, s) = \lambda$.
(3) If $\lambda \leq \mu$ then $\mathcal{I}_m(\lambda, r, s) \leq \mathcal{I}_m(\mu, r, s)$ and $\mathcal{C}_m(\lambda, r, s) \leq \mathcal{C}_m(\mu, r, s)$.
(4) $\mathcal{I}_m(\lambda \wedge \mu, r, s) = \mathcal{I}_m(\lambda, r, s) \wedge \mathcal{I}_m(\mu, r, s)$ and $\mathcal{C}_m(\lambda \vee \mu, r, s) = \mathcal{C}_m(\lambda, r, s) \vee \mathcal{C}_m(\mu, r, s)$.
(5) $\mathcal{I}_m(\mathcal{I}_m(\lambda, r, s), r, s) = \mathcal{I}_m(\lambda, r, s)$ and $\mathcal{C}_m(\mathcal{C}_m(\lambda, r, s), r, s) = \mathcal{C}_m(\lambda, r, s)$.
(6) $\underline{1} - \mathcal{C}_m(\lambda, r, s) = \mathcal{I}_m(\underline{1} - \lambda, r, s)$ and $\underline{1} - \mathcal{I}_m(\lambda, r, s) = \mathcal{C}_m(\underline{1} - \lambda, r, s)$

Definition 4 ([11]). *Let $(X, \mathcal{M}, \mathcal{M}^*)$ and $(Y, \mathcal{N}, \mathcal{N}^*)$ be two (r, s)-FMS's. Then $f : X \to Y$ is said to be fuzzy (r, s)-M-continuous if for every $\mu \in \mathcal{N}_{r,s}$, $f^{-1}(\mu) \in \mathcal{M}_{r,s}$.*

Theorem 2 ([11]). *Let $f : (X, \mathcal{M}, \mathcal{M}^*) \to (Y, \mathcal{N}, \mathcal{N}^*)$ be a function,*
(1) f is fuzzy (r, s)-M-continuous.
(2) $\underline{1} - f^{-1}(\mu) \in \mathcal{M}_{r,s}$, for each $\underline{1} - \mu \in \mathcal{N}_{r,s}$.
(3) $f(\mathcal{C}_m(\lambda, r, s)) \leq \mathcal{C}_m(f(\lambda), r, s)$, for $\lambda \in I^X$.
(4) $\mathcal{C}_m(f^{-1}(\mu), r, s)) \leq f^{-1}(\mathcal{C}_m(\mu, r, s))$, for $\mu \in I^Y$.
(5) $f^{-1}(\mathcal{I}_m(\mu, r, s)) \leq \mathcal{I}_m(f^{-1}(\mu), r, s)$, for $\mu \in I^Y$.
Then, $(1) \Leftrightarrow (2) \Rightarrow (3) \Leftrightarrow (4) \Leftrightarrow (5)$.

Definition 5 ([12,13]). *Let $(X, \mathcal{M}, \mathcal{M}^*)$ be an (r, s)-FMS and $\lambda \in I^X$. Then a fuzzy set λ is called a fuzzy (r, s)-minimal semiopen set [13] (respt., fuzzy (r, s)-minimal preopen set [12] in X if $\lambda \leq mC(mI(\lambda, r, s), r, s)$ (respt., $\lambda \leq mI(mC(\lambda, r, s), r, s)$. A fuzzy set λ is called a fuzzy (r, s)-minimal semiclosed (respt., fuzzy (r, s)-minimal preclosed set if the complement of λ is fuzzy (r, s)-minimal semiopen (respt., fuzzy (r, s)-minimal precopen).*

3 Fuzzy (r, s)-Minimal β-Open Sets

Definition 6. *Let $(X, \mathcal{M}, \mathcal{M}^*)$ be an (r, s)-FMS and $\lambda \in I^X$. Then a fuzzy set λ is said to be fuzzy (r, s)-minimal β-open in X if*

$$\lambda \leq mC(mI(mC(\lambda, r, s), r, s), r, s).$$

A fuzzy set λ is said to be fuzzy (r, s)-minimal β-closed if the complement of λ is fuzzy (r, s)-minimal β-open.

Remark 1. From definitions of fuzzy (r, s)-minimal semiopen, fuzzy (r, s)-minimal preopen and fuzzy (r, s)-minimal β-open sets, the following diagram is obtained but the converses is not be true in general as shown in the next example.

$$\begin{array}{ccccc} & & \text{fuzzy } (r,s)\text{-minimal semiopen} & & \\ & \nearrow & & \searrow & \\ \text{fuzzy } (r,s)\text{-minimal open} & & & & \text{fuzzy } (r,s)\text{-minimal } \beta\text{-open} \\ & \searrow & & \nearrow & \\ & & \text{fuzzy } (r,s)\text{-minimal preopen} & & \end{array}$$

Example 1. Let $X = I$. For $0 < n < 1$, consider the following fuzzy sets

$$\mu_n(x) = \begin{cases} \frac{x}{n} & \text{if } 0 < x \leq n \\ \frac{1-x}{1-n} & \text{if } n < x \leq 1; \end{cases}$$

$$\lambda(x) = \begin{cases} x + \frac{1}{2} & \text{if } 0 \leq x \leq \frac{1}{2} \\ -x + \frac{3}{2} & \text{if } \frac{1}{2} < x \leq 1. \end{cases}$$

$$\sigma(x) = \begin{cases} -\frac{x}{2} + \frac{1}{4} & \text{if } 0 \leq x \leq \frac{1}{2} \\ \frac{x}{2} - \frac{1}{4} & \text{if } \frac{1}{2} < x \leq 1. \end{cases}$$

(1) Define $\mathcal{M}, \mathcal{M}^* : I^X \to I$ on X by

$$\mathcal{M}(\mu) = \begin{cases} n & \text{if } \mu = \mu_n \\ \frac{7}{8} & \text{if } \mu = \sigma \\ 0 & \text{otherwise,} \end{cases}$$

$$\mathcal{M}^*(\mu) = \begin{cases} 1-n & \text{if } \mu = \mu_n \\ 0 & \text{if } \mu = \sigma \\ 1 & \text{otherwise.} \end{cases}$$

From $\mathcal{C}_m(\mathcal{I}_m(\lambda, \frac{3}{4}, \frac{1}{4}), \frac{3}{4}, \frac{1}{4}) = \underline{1} - \sigma \geq \lambda$, λ is fuzzy $(\frac{3}{4}, \frac{1}{4})$-minimal β-open but it is not fuzzy $(\frac{3}{4}, \frac{1}{4})$-minimal preopen.

(2) Define $\mathcal{M}, \mathcal{M}^* : I^X \to I$ on X by

$$\mathcal{M}(\mu) = \begin{cases} n & \text{if } \mu = \mu_n \\ \frac{7}{8} & \text{if } \mu = \underline{1} - \lambda \\ 0 & \text{otherwise,} \end{cases}$$

$$\mathcal{M}^*(\mu) = \begin{cases} 1-n & \text{if } \mu = \mu_n \\ 0 & \text{if } \mu = \underline{1} - \lambda \\ 1 & \text{otherwise.} \end{cases}$$

From $\mathcal{C}_m(\mathcal{I}_m(\sigma, \frac{3}{4}, \frac{1}{4}), \frac{3}{4}, \frac{1}{4}) = \underline{1} - \lambda \geq \sigma$ and $\mathcal{I}_m(\sigma, \frac{3}{4}, \frac{1}{4}) = \underline{0}$, σ is fuzzy $(\frac{3}{4}, \frac{1}{4})$-minimal β-open but it is not fuzzy $(\frac{3}{4}, \frac{1}{4})$-minimal semiopen.

Lemma 1. *Let $(X, \mathcal{M}, \mathcal{M}^*)$ be an (r,s)-FMS and $\lambda \in I^X$. Then a fuzzy set λ is fuzzy (r,s)-minimal β-closed if and only if $mI(mC(mI(\lambda, r, s), r, s), r, s) \leq \lambda$.*

Theorem 3. *Let $(X, \mathcal{M}, \mathcal{M}^*)$ be an (r,s)-FMS. Any union of fuzzy (r,s)-minimal β-open sets is fuzzy (r,s)-minimal β-open.*

Proof. Let λ_i be a fuzzy (r,s)-minimal β-open set for $i \in J$. Then from $\lambda_i \leq \vee\lambda_i$ for $i \in J$,

$$\lambda_i \leq mC(mI(mC(\lambda_i,r,s),r,s),r,s) \leq mC(mI(mC(\vee\lambda_i,r,s),r,s),r,s).$$

This implies $\vee\lambda_i \leq mC(mI(mC(\vee\lambda_i,r,s),r,s),r,s)$ and so $\vee\lambda_i$ is fuzzy (r,s)-minimal β-open. □

Remark 2. In general, the intersection of two fuzzy (r,s)-minimal β-open sets may not be fuzzy (r,s)-minimal β-open as shown in the next example.

Example 2. Let $X = I$. For $0 < n < 1$, consider the following fuzzy sets

$$\mu_n(x) = \begin{cases} \frac{x}{n} & \text{if } 0 < x \leq n \\ \frac{1-x}{1-n} & \text{if } n < x \leq 1; \end{cases}$$

Define $\mathcal{M}, \mathcal{M}^* : I^X \to I$ on X by

$$\mathcal{M}(\mu) = \begin{cases} n & \text{if } \mu = \mu_n \\ 0 & \text{otherwise,} \end{cases}$$

$$\mathcal{M}^*(\mu) = \begin{cases} 1-n & \text{if } \mu = \mu_n \\ 1 & \text{otherwise.} \end{cases}$$

Consider a $(\frac{3}{4},\frac{1}{4})$-fuzzy minimal structure $(\mathcal{M}_{\frac{3}{4}}, \mathcal{M}^*_{\frac{1}{4}})$ as follows

$$\mathcal{M}_{\frac{3}{4},\frac{1}{4}} = \{\mu \in I^X \mid \mathcal{M}(\mu) \geq \frac{3}{4} \text{ and } \mathcal{M}^*(\mu) \leq \frac{1}{4}\}.$$

Let us consider two fuzzy sets λ and γ:

$$\lambda = \mu_{\frac{3}{4}} \quad \gamma = \mu_{\frac{7}{8}}$$

Then obviously λ and γ are fuzzy $(\frac{3}{4},\frac{1}{4})$-minimal β-open. But since $mI(\lambda\wedge\gamma,\frac{3}{4},\frac{1}{4}) = \underline{0}$, $\lambda \wedge \gamma$ can not be fuzzy $(\frac{3}{4},\frac{1}{4})$-minimal β-open.

Definition 7. *Let $(X,\mathcal{M},\mathcal{M}^*)$ be an (r,s)-FMS. For $\lambda \in I^X$, $m\beta C(\lambda,r,s)$ and $m\beta I(\lambda,r)$, respectively, are defined as the following:*

$$m\beta C(\lambda,r,s) = \wedge\{\gamma \in I^X : \lambda \leq \gamma,\ \ \gamma \textit{ is fuzzy } (r,s)\textit{-minimal } \beta\textit{-closed}\};$$

$$m\beta I(\lambda,r,s) = \vee\{\mu \in I^X : \mu \leq \lambda,\ \ \mu \textit{ is fuzzy } (r,s)\textit{-minimal } \beta\textit{-open}\}.$$

Theorem 4. *Let $(X,\mathcal{M},\mathcal{M}^*)$ be an (r,s)-FMS and $\lambda \in I^X$. Then*
(1) $m\beta I(\lambda,r,s) \leq \lambda$.
(2) If $\lambda \leq \mu$, then $m\beta I(\lambda,r,s) \leq m\beta I(\mu,r,s)$.
(3) λ is (r,s)-minimal β-open iff $m\beta I(\lambda,r,s) = \lambda$.
(4) $m\beta I(\beta mI(\lambda,r,s),r,s) = m\beta I(\lambda,r,s)$.
(5) $m\beta C(\underline{1}-\lambda,r,s) = \underline{1}-m\beta I(\lambda,r,s)$ and $m\beta I(\underline{1}-\lambda,r,s) = \underline{1}-m\beta C(\lambda,r,s)$.

Proof. (1), (2), (3) and (4) Obvious.
(5) For $\lambda \in I^X$,

$$\begin{aligned}\underline{1} - m\beta I(\lambda, r, s) &= \underline{1} - \vee\{\mu \in I^X : \mu \leq \lambda, \ \ \mu \text{ is fuzzy } (r,s)\text{-minimal } \beta\text{-open}\} \\ &= \wedge\{\underline{1} - \mu : \mu \leq \lambda, \ \ \mu \text{ is fuzzy } (r,s)\text{-minimal } \beta\text{-open}\} \\ &= \wedge\{\underline{1} - \mu : \underline{1} - \lambda \leq \underline{1} - \mu, \ \ \mu \text{ is fuzzy } (r,s)\text{-minimal } \beta\text{-open}\} \\ &= m\beta C(\underline{1} - \lambda, r, s).\end{aligned}$$

Similarly, we can show that $m\beta I(\underline{1} - \lambda, r, s) = \underline{1} - m\beta C(\lambda, r, s)$. □

Theorem 5. *Let $(X, \mathcal{M}, \mathcal{M}^*)$ be an (r, s)-FMS and $\lambda \in I^X$. Then*
(1) $\lambda \leq m\beta C(\lambda, r, s)$.
(2) If $\lambda \leq \mu$, then $m\beta C(\lambda, r, s) \leq m\beta C(\mu, r, s)$.
(3) λ is r, s-minimal β-closed iff $m\beta C(\lambda, r, s) = \lambda$.
(4) $m\beta C(m\beta C(\lambda, r, s), r, s) = m\beta C(\lambda, r, s)$.

Proof. It is similar to the proof of Theorem 4. □

4 Fuzzy (r, s)-M β-Continuous Mappings, Fuzzy (r, s)-$M(M^*)$ β-Open Mappings

In this section, we introduce the concepts of fuzzy (r, s)-M β-continuous mapping, fuzzy (r, s)-M β-open mapping and fuzzy (r, s)-M^* β-open mapping, and investigate characterization for such mappings.

Definition 8. *Let $(X, \mathcal{M}, \mathcal{M}^*)$ and $(Y, \mathcal{N}, \mathcal{N}^*)$ be (r, s)-FMS's. Then a mapping $f : X \to Y$ is said to be fuzzy (r, s)-M β-continuous if $f^{-1}(\lambda)$ is a fuzzy (r, s)-minimal β-open set for each fuzzy (r, s)-minimal open set $\lambda \in I^Y$.*

Let $(X, \mathcal{M}, \mathcal{M}^*)$ and $(Y, \mathcal{N}, \mathcal{N}^*)$ be (r, s)-FMS's. Then a mapping $f : X \to Y$ is said to be *fuzzy (r, s)-M semicontinuous* [13] (respt., *fuzzy (r, s)-M precontinuous* [12]) if $f^{-1}(\lambda)$ is a fuzzy (r, s)-minimal semiopen set (respt., fuzzy (r, s)-minimal semiopen set) for each fuzzy (r, s)-minimal open set $\lambda \in I^Y$.

Remark 3. From Remark 1, we can easily show that every fuzzy (r, s)-M semicontinuous mapping(fuzzy (r, s)-M precontinuous mapping) is fuzzy (r, s)-M β-continuous but the converse is not true in general.

fuzzy (r, s)-M semicontinuous
$\nearrow$ $\searrow$
fuzzy (r, s)-M continuous fuzzy (r, s)-M β-continuous
$\searrow$ $\nearrow$
fuzzy (r, s)-M precontinuous

Theorem 6. *Let $f : X \to Y$ be a mapping on (r, s)-FMS's $(X, \mathcal{M}, \mathcal{M}^*)$ and $(Y, \mathcal{N}, \mathcal{N}^*)$. Then the following statements are equivalent:*
(1) f is fuzzy (r, s)-M β-continuous.

(2) $f^{-1}(\mu)$ is a fuzzy (r, s)-minimal β-closed set for each fuzzy (r, s)-minimal closed set $\mu \in I^Y$.
(3) $f(m\beta C(\lambda, r, s)) \subseteq mC(f(\lambda), r, s)$ for $\lambda \in I^X$.
(4) $m\beta C(f^{-1}(\mu), r, s) \subseteq f^{-1}(mC(\mu, r, s))$ for $\mu \in I^Y$.
(5) $f^{-1}(mI(\mu, r, s)) \subseteq m\beta I(f^{-1}(\mu), r, s)$ for $\mu \in I^Y$.

Proof. (1) $\Rightarrow$ (2) Obvious.
(2) $\Rightarrow$ (3) For $\lambda \in I^X$,

$$\begin{aligned} f^{-1}(mC(f(\lambda), r, s)) &= f^{-1}(\wedge\{\gamma \in I^Y : f(\lambda) \leq \gamma \text{ and } \gamma \text{ is fuzzy } (r, s)\text{-minimal closed}\}) \\ &= \wedge\{f^{-1}(\gamma) \in I^X : \lambda \leq f^{-1}(\gamma) \text{ and} \\ &\qquad f^{-1}(\gamma) \text{ is fuzzy } (r, s)\text{-minimal } \beta\text{-closed}\} \\ &\geq \wedge\{\sigma \in I^X : \lambda \leq \sigma \text{ and } \sigma \text{ is fuzzy } (r, s)\text{-minimal } \beta\text{-closed}\} \\ &= m\beta C(\lambda, r, s). \end{aligned}$$

This implies $f(m\beta C(\lambda, r, s)) \leq mC(f(\lambda), r, s)$.
(3) $\Rightarrow$ (4) For $\mu \in I^Y$,

$$f(m\beta C(f^{-1}(\mu), r, s)) \leq mC(f(f^{-1}(\mu)), r, s) \leq mC(\mu, r, s).$$

So $m\beta C(f^{-1}(\mu), r, s) \leq f^{-1}(mC(\mu, r, s))$.
(4) $\Rightarrow$ (5) For $\mu \leq Y$, from Theorem 4, it follows

$$\begin{aligned} f^{-1}(mI(\mu, r, s)) &= f^{-1}(\underline{1} - mC(\underline{1} - \mu, r, s)) \\ &= \underline{1} - f^{-1}(mC(\underline{1} - \mu, r, s)) \\ &\leq \underline{1} - m\beta C(f^{-1}(\underline{1} - \mu), r, s) \\ &= m\beta I(f^{-1}(\mu), r, s). \end{aligned}$$

This implies $f^{-1}(mI(\mu, r, s)) \leq m\beta I(f^{-1}(\mu), r, s)$.
(5) $\Rightarrow$ (1) Let $\lambda \in I^Y$ be any fuzzy (r, s)-minimal open set. By hypothesis, $f^{-1}(\lambda) = f^{-1}(mI(\lambda, r, s)) \leq m\beta I(f^{-1}(\lambda), r, s)$. From Theorem 4, $f^{-1}(\lambda) = m\beta I(f^{-1}(\lambda), r, s)$ and $f^{-1}(\lambda)$ is fuzzy (r, s)-minimal β-open. So f is fuzzy (r, s) M β-continuous. □

Definition 9. *Let $(X, \mathcal{M}, \mathcal{M}^*)$ and $(Y, \mathcal{N}, \mathcal{N}^*)$ be r-FMS's. Then a mapping $f : X \to Y$ is said to be fuzzy (r, s)-M^*-β-open if for every fuzzy (r, s)-minimal β-open set $\lambda \in I^X$, $f(\lambda)$ is fuzzy (r, s)-minimal open.*

Theorem 7. *Let $f : X \to Y$ be a mapping on (r, s)-FMS's $(X, \mathcal{M}, \mathcal{M}^*)$ and $(Y, \mathcal{N}, \mathcal{N}^*)$.*
(1) f is fuzzy (r, s)-M^*-β-open.
(2) $f(m\beta I(\lambda, r, s)) \leq mI(f(\lambda), r, s)$ for $\lambda \in I^X$.
(3) $m\beta I(f^{-1}(\gamma), r, s) \leq f^{-1}(mI(\gamma, r, s))$ for $\gamma \in I^Y$.
Then (1) $\Rightarrow$ (2) $\Leftrightarrow$ (3).

Proof. (1) $\Rightarrow$ (2) For $\lambda \in I^X$,

$$\begin{aligned} f(m\beta I(\lambda, r, s)) &= f(\vee\{\mu \in I^X : \mu \leq \lambda,\ \mu \text{ is fuzzy } (r, s)\text{-minimal } \beta\text{-open}\}) \\ &= \vee\{f(\mu) \in I^Y : f(\mu) \leq f(\lambda),\ f(\mu) \text{ is fuzzy } (r, s)\text{-minimal open}\} \\ &\leq \vee\{\sigma \in I^Y : \sigma \leq f(\lambda),\ \sigma \text{ is fuzzy } (r, s)\text{-minimal open}\} \\ &= mI(f(\lambda), r, s). \end{aligned}$$

Hence $f(m\beta I(\lambda,r,s)) \leq mI(f(\lambda),r,s)$.
(2) ⇒ (3)
For $\gamma \in I^Y$, from (3),

$$f(m\beta I(f^{-1}(\gamma),r,s)) \leq mI(f(f^{-1}(\gamma)),r,s) \leq mI(\gamma,r,s).$$

Similarly, we have the implication (3) ⇒ (2).

□

Remark 4. From (1) of Theorem 1, we can show that the implication (2) ⇒ (1) in Theorem 7 is not always true.

Definition 10 ([11]). *Let X be a nonempty set and $\mathcal{M}, \mathcal{M}^* : I^X \to I$. Then the pair $(\mathcal{M}, \mathcal{M}^*)$ is said to have the property (P) if for $\lambda_j \in I^X$, $j \in J$*

$$\mathcal{M}(\bigvee_{j\in J} \lambda_j) \geq \bigwedge_{j\in J} \mathcal{M}(\lambda_j) \text{ and } \mathcal{M}^*(\bigvee_{j\in J} \lambda_j) \leq \bigvee_{j\in J} \mathcal{M}^*(\lambda_j).$$

Theorem 8 ([11]). *Let $(X, \mathcal{M}, \mathcal{M}^*)$ be an (r,s)-FMS with the property (P). Then for $\lambda \in I^X$, $\mathcal{I}_m(\lambda,r,s) = \lambda$ iff λ is fuzzy (r,s)-minimal open.*

From the above Theorem 8, obviously the following corollary is obtained:

Corollary 1. *Let $f : X \to Y$ be a mapping on (r,s)-FMS's $(X, \mathcal{M}, \mathcal{M}^*)$ and $(Y, \mathcal{N}, \mathcal{N}^*)$. If $(Y, \mathcal{N}, \mathcal{N}^*)$ has the property (P), then the following are equivalent:*
(1) f is fuzzy (r,s)-M^*-β-open.
(2) $f(m\beta I(\lambda,r,s)) \leq mI(f(\lambda),r,s)$ for $\lambda \in I^X$.
(3) $m\beta I(f^{-1}(\mu),r,s) \leq f^{-1}(mI(\mu,r,s))$ for $\mu \in I^Y$.

Definition 11. *Let $(X, \mathcal{M}, \mathcal{M}^*)$ and $(Y, \mathcal{N}, \mathcal{N}^*)$ be r-FMS's. Then a mapping $f : X \to Y$ is said to be fuzzy (r,s)-M-β-open if for fuzzy (r,s)-minimal open set $\lambda \in I^X$, $f(\lambda)$ is fuzzy (r,s)-minimal β-open.*

Theorem 9. *Let $f : X \to Y$ be a mapping on (r,s)-FMS's $(X, \mathcal{M}, \mathcal{M}^*)$ and $(Y, \mathcal{N}, \mathcal{N}^*)$. Then the following are equivalent:*
(1) f is fuzzy (r,s)-M-β-open.
(2) $f(mI(\lambda,r,s)) \leq m\beta I(f(\lambda),r,s)$ for $\lambda \in I^X$.
(3) $mI(f^{-1}(\mu),r,s) \leq f^{-1}(m\beta I(\mu,r,s))$ for $\mu \in I^Y$.

Proof. (1) ⇒ (2) For $\lambda \in I^X$,

$$\begin{aligned} f(mI(\lambda,r,s)) &= f(\vee\{\sigma \in I^X : \sigma \leq \lambda,\ \sigma \text{ is fuzzy } (r,s)\text{-minimal open}\}) \\ &= \vee\{f(\sigma) \in I^Y : f(\sigma) \leq f(\lambda),\ f(\sigma) \text{ is fuzzy } (r,s)\text{-minimal } \beta\text{-open}\} \\ &\leq \vee\{\gamma \in I^X : \gamma \leq f(\lambda),\ \gamma \text{ is fuzzy } (r,s)\text{-minimal } \beta\text{-open}\} \\ &= m\beta I(f(\lambda),r,s). \end{aligned}$$

Hence $f(mI(\lambda,r,s)) \leq m\beta I(f(\lambda),r,s)$.
(2) ⇒ (3) For $\mu \in I^Y$, $f(mI(f^{-1}(\mu),r,s)) \leq m\beta I(f(f^{-1}(\mu)),r,s) \leq m\beta I(\mu,r,s)$. Hence (3) is obtained.

(3) $\Rightarrow$ (2) Obvious.

(2) $\Rightarrow$ (1) Let $\lambda \in I^X$ be a fuzzy (r, s)-minimal open set. Then $\lambda = mI(\lambda, r, s)$ and by (2), $f(\lambda) = m\beta I(f(\lambda), r, s)$ and $f(\lambda)$ is fuzzy (r, s)-minimal β-open. Hence f is a fuzzy (r, s)-M-β-open mapping.

□

References

1. Atanassov, K.: Intuitionistic fuzzy sets. Fuzzy Sets and Systems 20(1), 87–96 (1986)
2. Chang, C.L.: Fuzzy topological spaces. J. Math. Anal. Appl. 24, 182–190 (1968)
3. Chattopadhyay, K.C., Hazra, R.N., Samanta, S.K.: Gradation of openess: fuzzy topology. Fuzzy Sets and Systems 49, 237–242 (1992)
4. Çoker: An introduction to intuitionistic fuzzy topological spaces. Fuzzy Sets and Systems 88, 81–89 (1997)
5. Çoker, D., Dimirci, M.: An introduction to intuitionistic fuzzy topological spaces in Šostak sense. Buseful 67, 67–76 (1996)
6. Höhle, U.: Upper semicontinuous fuzzy sets and applications. J. Math. Anall. Appl. 78, 659–673 (1980)
7. Höhle, U., Šostak, A.P.: A general theory of fuzzy topological spaces. Fuzzy Sets and Systems 73, 131–149 (1995)
8. Höhle, U., Šostak, A.P.: Axiomatic Foundations of Fixed-Basis fuzzy topology, ch. 3. The Handbooks of Fuzzy sets series, vol. 3. Kluwer Academic Publishers, Dordrecht (1999)
9. Kubiak, T.: On fuzzy topologies. Ph. D. Thesis, A. Mickiewicz, Poznan (1985)
10. Kubiak, T., Šostak, A.P.: Lower set-valued fuzzy topologies. Quaestions Math. 20(3), 423–429 (1997)
11. Min, W.K., Abbas, S.E.: (r, s)-fuzzy minimal structures and (r, s)-fuzzy minimal spaces. Journal of Intelligent and Fuzzy Systems, doi:10.3233/IFS-120669 (accepted)
12. Min, W.K., Kim, Y.K.: Fuzzy (r, s)-minimal preopen sets and fuzzy (r, s)-M precontinuous mappings on (r, s)-fuzzy minimal spaces (submitted)
13. Min, W.K., Kim, Y.K.: Fuzzy (r, s)-minimal semiopen sets and fuzzy (r, s)-M semicontinuous mappings on (r, s)-fuzzy minimal spaces (submitted)
14. Samanta, S.K., Mondal, T.K.: On intuitionistic gradation of openness. Fuzzy Sets and Systems 131, 323–336 (2002)
15. Šostak, A.P.: On a fuzzy topological structure, Suppl. Rend. Circ. Matem. Palerms Ser. II 11, 89–103 (1985)
16. Zadeh, L.A.: Fuzzy sets. Information and Control 8, 338–353 (1965)

Handling Very Large Cooccurrence Matrices in Fuzzy Co-clustering by Sampling Approaches

Katsuhiro Honda[1], Akira Notsu[1], and Chi-Hyon Oh[2]

[1] Graduate School of Engineering, Osaka Prefecture University,
1-1 Gakuen-cho, Nakaku, Sakai, Osaka, 599-8531, Japan
{honda,notsu}@cs.osakafu-u.ac.jp
[2] Faculty of Liberal Arts and Sciences, Osaka University of Economics and Law,
6-10, Gakuonji, Yao, Osaka, 581-8511 Japan
ch_oh@keiho-u.ac.jp

Abstract. Handling very large data sets is a significant issue in many applications of data analysis. In Fuzzy *c*-Means (FCM), several sampling approaches for handling very large data have been proved to be useful. In this paper, the sampling approaches are applied to fuzzy co-clustering tasks for handling cooccurrence matrices composed of many objects. The goal of co-clustering is simultaneously partition both objects and items into co-clusters and item memberships are used for characterizing each co-cluster instead of cluster centers in the conventional FCM. In some modified approaches, item memberships are utilized in conjunction with other objects for inheriting the property of other sample sets.

Keywords: fuzzy clustering, co-clustering, very large data set.

1 Introduction

In many data analysis applications, we often have very large data sets and suffer from computational overloads. Havens *et al.* [1] proposed several sampling approaches for handling very large data sets composed of a number of objects in Fuzzy *c*-Means (FCM) [2]. The simplest approach is to apply the conventional FCM algorithm only to a subset of the data set [3], which is called *random sample* and *extend* approach (rseFCM), although the clustering quality might be significantly degraded. Then, the sampling approach was modified for efficiently applying the FCM algorithm to all subsets of the whole data set. In the *single pass* approach (spFCM) [4], the weighted FCM algorithm is sequentially performed to the subsets, where the cluster centers of the previous sub-process (with previous subset) are added to the present subset with a certain weights in each sub-process for inheriting the current cluster structures to the next sub-process. Beside such sequential approach, the *online* approach (oFCM) [5] perform the conventional FCM with each of all subsets in parallel, and re-perform the weighted FCM algorithm with the derived cluster centers in conjunction with their membership weights.

Y.I. Cho et al. (eds.), *Soft Computing in Artificial Intelligence*,
Advances in Intelligent Systems and Computing 270,
DOI: 10.1007/978-3-319-05515-2_3, © Springer International Publishing Switzerland 2014

In this paper, the sampling approaches are applied to fuzzy co-clustering in order to estimate co-cluster structures of very large cooccurrence matrices including a number of objects. The goal of co-clustering is to extract co-clusters composed of object-item pairs, which are mutually related. The data to be clustered is given by a cooccurrence matrix, whose elements represent the degree of coocurrence of object-item pairs, e.g., the number of appearance of keywords in documents and the frequencies of purchase of items by customers.

Fuzzy clustering for categorical multivariate data (FCCM) [6] is an FCM-type co-clustering model, in which fuzzy partition of both objects and items are estimated based on the FCM-like concept. The clustering criterion is given by the degree of aggregation to be maximized while different constraints are forced to the two memberships of objects and items with the goal of avoiding a trivial solution, i.e., the sum of memberships of items are forced to be 1 in each cluster and the memberships represent the relative responsibility of items in each cluster. Then, the item memberships can be used for characterizing each cluster in a same manner with the cluster centers in FCM. Then, the proposed sampling approach performs FCCM by using item memberships for inheriting the current co-cluster structures.

The remaining part of this paper is organized as follows: Section 2 gives a brief review on FCCM and the new sampling approaches are proposed in Section 3. Several experimental results are shown in Section 4 and a summary conclusion is presented in Section 5.

2 A Brief Review on FCM-Type Co-clustering

Co-clustering is a technique for estimating the intrinsic cluster structures from co-occurrence information among objects and items. Assume that we have a co-occurrence matrix $R = \{r_{ij}\}$ on objects $i = 1,\dots,n$ and items $i = 1,\dots,m$, in which r_{ij} represent the degree of occurrence of item *j* with object *i*. Oh *et al.* [6] proposed the FCM-type co-clustering model, which is called FCCM, by modifying the FCM algorithm for handling cooccurrence information, where the cluster aggregation degree of each cluster is maximized.

$$L_{fccm} = \sum_{c=1}^{C}\sum_{i=1}^{n}\sum_{j=1}^{m} u_{ci} w_{cj} r_{ij} + \lambda_u \sum_{c=1}^{C}\sum_{i=1}^{n} u_{ci} \log u_{ci} + \lambda_w \sum_{c=1}^{C}\sum_{j=1}^{m} w_{cj} \log w_{cj} \tag{1}$$

where u_{ci} and w_{cj} are the membership degree of user *i* and item *j* in cluster *c*. *C* is the number of clusters. The sum of memberships are constrained as $\sum_{c=1}^{C} u_{ci} = 1$ and $\sum_{j=1}^{m} w_{cj} = 1$, respectively. The entropy terms are the fuzzification penalty in the entropy-based fuzzification approach [7], in which larger λ brings a fuzzier partition. Following $\sum_{c=1}^{C} u_{ci} = 1$, each object tend to belong to a solo cluster, i.e., they are forced a exclusive partition. On the other hand, $\sum_{j=1}^{m} w_{cj} = 1$ implies that the item memberships represent the mutual responsibility of them in cluster c. So, the item memberships are available in characterizing each cluster in the same manner with the cluster centers of FCM while each item can belong to multiple clusters (or be rejected by all clusters).

The clustering algorithm is composed of the FCM-like iterative process of updating object and item memberships until convergent. It has been shown that the FCCM algorithm is useful in various applications such as collaborative filtering used in conjunction with exclusive partitions [8] and some cluster validity measures [9].

3 Modified Sampling Approaches for Fuzzy Co-clustering

In this paper, the three sampling approaches for FCM introduced by Havens *et al.* [1] are modified with the goal of applying to fuzzy co-clustering. Here, we should note that the sampling approaches must be utilized in a slightly different manner from the FCM context. In the FCM context, cluster centers estimated from a subset of objects are used for characterizing each cluster and they are identified with representative objects in sequential process. On the other hand, in the FCCM context, we do NOT have such representative prototype but have item memberships, which merely connect familiar items to cluster objects. In the followings, the applicability of item memberships for representative cluster prototypes is discussed.

3.1 Random Sample and Extend Approach: rseFCCM

In the random sample and extend approach, the conventional FCCM algorithm is applied only to a randomly extracted subset composed of n_s objects, where the subset constructs an $n_s \times m$ cooccurrence matrix R_s.

Once the memberships for the subset are estimated, we can calculate the memberships of the remaining objects by using the item memberships, which are already estimated.

3.2 Single-Pass Approach: spFCCM

In the single-pass approach, before applying the FCCM algorithm, the whole data set is divided into s subsets, each of which is composed of randomly selected n_s objects. Then, s subsets are represented by $n_s \times m$ cooccurrence matrices $R_1, \cdots, R_s$, and the FCCM algorithm is sequentially performed with the subsets.

First, the conventional FCCM algorithm is applied to R_1. Then, in the remaining process, $R_2, \cdots, R_s$ are sequentially analyzed by the weighted FCCM algorithm, in which the previous item memberships are added to the subset in conjunction with their weights, i.e., R_k is $(n_s + C) \times m$ matrix, whose $(n_s + t)$ th row is $(w_{t1} \ \cdots \ w_{tm})$ of the previous process.

The weighted FCCM, which is newly considered in this paper, is a modified version of FCCM for handling data with responsibility weights. Assume that the responsibility weight of object i is given as α_i and we should extract clusters reflecting the object responsibilities. Considering the object weights, the objective function of FCCM is modified as follows:

$$L_{wfccm} = \sum_{c=1}^{C}\sum_{i=1}^{n}\sum_{j=1}^{m} \alpha_i u_{ci} w_{cj} r_{ij} + \lambda_u \sum_{c=1}^{C}\sum_{i=1}^{n} u_{ci} \log u_{ci} + \lambda_w \sum_{c=1}^{C}\sum_{j=1}^{m} w_{cj} \log w_{cj} \tag{2}$$

where u_{ci} and w_{cj} follow the same constraint with the conventional FCCM.

Considering the optimality conditions, the updating rules for u_{ci} and w_{cj} are given as follows:

$$u_{ci} = \frac{\exp\left(\lambda_u^{-1} \sum_{j=1}^{m} w_{cj} r_{ij}\right)}{\sum_{l=1}^{C} \exp\left(\lambda_u^{-1} \sum_{j=1}^{m} w_{lj} r_{ij}\right)} \tag{3}$$

$$w_{cj} = \frac{\exp\left(\lambda_w^{-1} \sum_{i=1}^{n} \alpha_i u_{ci} r_{ij}\right)}{\sum_{l=1}^{m} \exp\left(\lambda_w^{-1} \sum_{i=1}^{n} \alpha_i u_{ci} r_{il}\right)} \tag{4}$$

Here, the updating rules are reduced to the conventional ones of FCCM when all $\alpha_i = 1$.

In spFCCM, $\alpha_i = 1$ for $i = 1, \cdots, n_s$ while $\alpha_i = \sum_{i=1}^{n_s+C} u_{ci} \alpha_i$ for $i = n_s + 1, \cdots, n_s + C$. In order to efficiently inherit the previous co-cluster structures, the initial memberships in each sub-process are given by the previous results. After finishing the all sub-process, the memberships of objects for $R_1, \cdots, R_{s-1}$ should be recalculated using the final item memberships.

3.3 Online Approach: oFCCM

In the online approach, before applying the FCCM algorithm, the whole data set is also divided into s subsets, each of which is composed of randomly selected n_s objects. Then, s subsets are represented by $n_s \times m$ cooccurrence matrices $R_1, \cdots, R_s$, and the FCCM algorithm is performed with each of subsets in parallel, i.e., we have $s \times C$ item membership vectors $\mathbf{w}_c^k = \left(w_{c1}^k \;\; \cdots \;\; w_{cm}^k\right), k = 1, \cdots, s, c = 1, \cdots, C$.

After finishing the all parallel processes, the item membership vectors are gathered into an $(s\times C)\times m$ cooccurrence matrix $\tilde{R}$ and the weighted FCCM algorithm is performed with weights $\alpha_i^k = \sum_{i=1}^{n_s} u_{ci}^k$.

In the same manner with spFCCM, once we derive the final item memberships, the memberships of each object should be finalized by recalculation.

4 Numerical Experiments

A comparative experiment was performed with an artificially generated cooccurrence matrix. Figure 1(a) shows the base data matrix composed of 100 objects and 60 items, which includes four co-clusters. Black and white cells indicate $r_{ij}=1$ and $r_{ij}=0$, respectively. Note that objects are exclusively assigned to a solo cluster while some items are shared by multiple clusters. Especially, the last two items are shared by all four clusters. 10 noisy subsets, each of which is composed of 100 objects an 60 items, were generated from Fig. 1(a): `1' elements and `0' elements were randomly replaces with `0' and `1' with the probability of 50% and 10%, respectively. Then, the 10 subsets $R_1,\cdots,R_{10}$ were gathered into a 1000×60 cooccurrence matrix R. For example, the first subset R_1 is shown in Fig. 1(b).

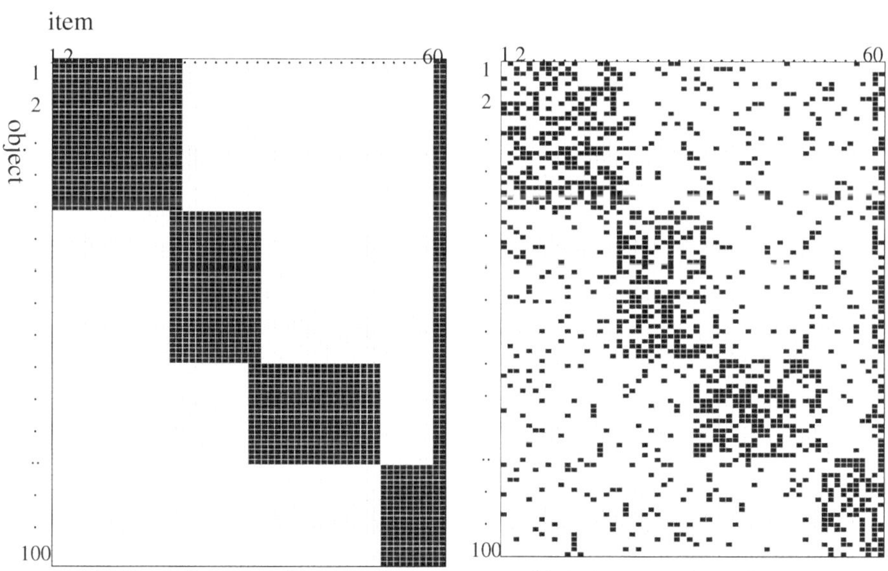

(a) Base data set without noise (b) Subset R_1 with noise

Fig. 1. The artificially generated cooccurrence matrices. Black and white cells indicate $r_{ij}=1$ and $r_{ij}=0$, respectively.

First, the conventional FCCM algorithm with $C = 4$, $\lambda_u = 0.05$ and $\lambda_w = 5.0$ was applied to the 100×60 base matrix of Fig. 1(a) and the item memberships are given as shown in Fig. 2. The membership degrees are depicted by color gradation, i.e., each row represent the membership vector in each cluster $\boldsymbol{w}_c = (w_{c1} \ \cdots \ w_{cm})$, where a high depth means a large membership. The FCCM algorithm is useful for revealing intrinsic co-cluster structures by properly estimating the memberships of object and items simultaneously.

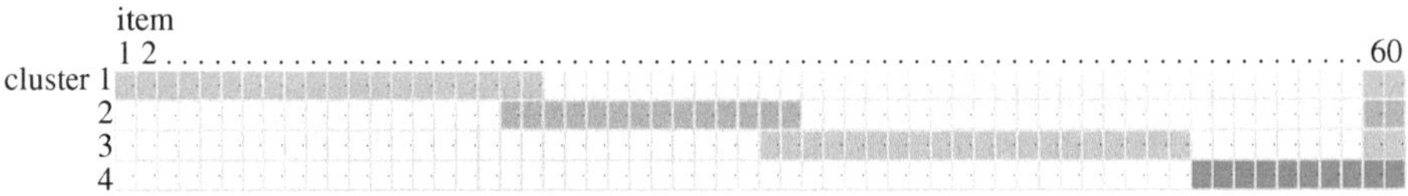

Fig. 2. Item memberships without noise. (Membership degrees are depicted by color gradation.)

Second, the FCCM algorithm with $C = 4$, $\lambda_u = 0.001$ and $\lambda_w = 100.0$ was applied to the whole data matrix R composed of 1000 objects. Figure 3 shows the derived item membership of each cluster. Here, the average correlation coefficient among Figs. 2 and 3 is 0.985. Even though the data matrix includes noise, the FCCM algorithm could successfully reveal the intrinsic co-cluster structure with a sufficient number of objects.

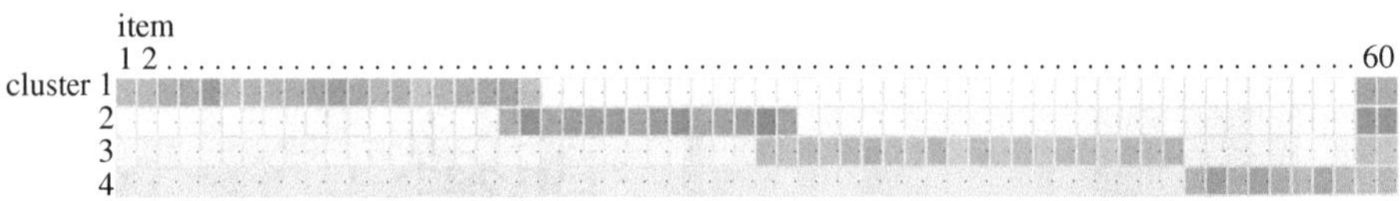

Fig. 3. Item memberships derived from whole data matrix R (1000×60). (Membership degrees are depicted by color gradation.)

Third, the rseFCCM algorithm was applied with $C = 4$, $\lambda_u = 0.001$ and $\lambda_w = 100.0$. In general, the FCCM algorithm is performed only once with one of subsets. However, in this experiment, the FCCM algorithm was independently applied to all 10 subsets $R_1, \cdots, R_{10}$ for discussing the stability against the influences of random selection. Figure 4 compares the item membership vectors given by the 10 subsets. The average correlation coefficient with Fig. 2 was 0.910. Because we had only an insufficient number of objects in each trial, the correlation coefficient was degenerated from that of the whole data case of Fig. 3.

Fourth, the spFCCM algorithm was applied with $C = 4$, $\lambda_u = 0.001$ and $\lambda_w = 100.0$. In spFCCM, the weighted FCCM algorithm was sequentially applied to $R_1, \cdots, R_{10}$, and then, the item memberships shown in Fig. 5 was derived in the final process. The average correlation coefficient was 0.912. The result of rseFCCM was slightly improved by spFCCM because all subsets were sequentially reflected in this approach. However, the result is not similar enough to that of the whole data analysis, i.e., Fig. 3.

In spFCCM, the memberships for the inherited objects (item membership of previous sub-process) were gradually `significantly fuzzified' and became $u_{ci} \to 1/C$ for all clusters in the final sub-process. This ambiguity might bring only a little improvement and the result implies that we should also consider the membership classification in the sequential process although it is not needed in the FCCM case [1]. This may be a special feature of the FCCM case.

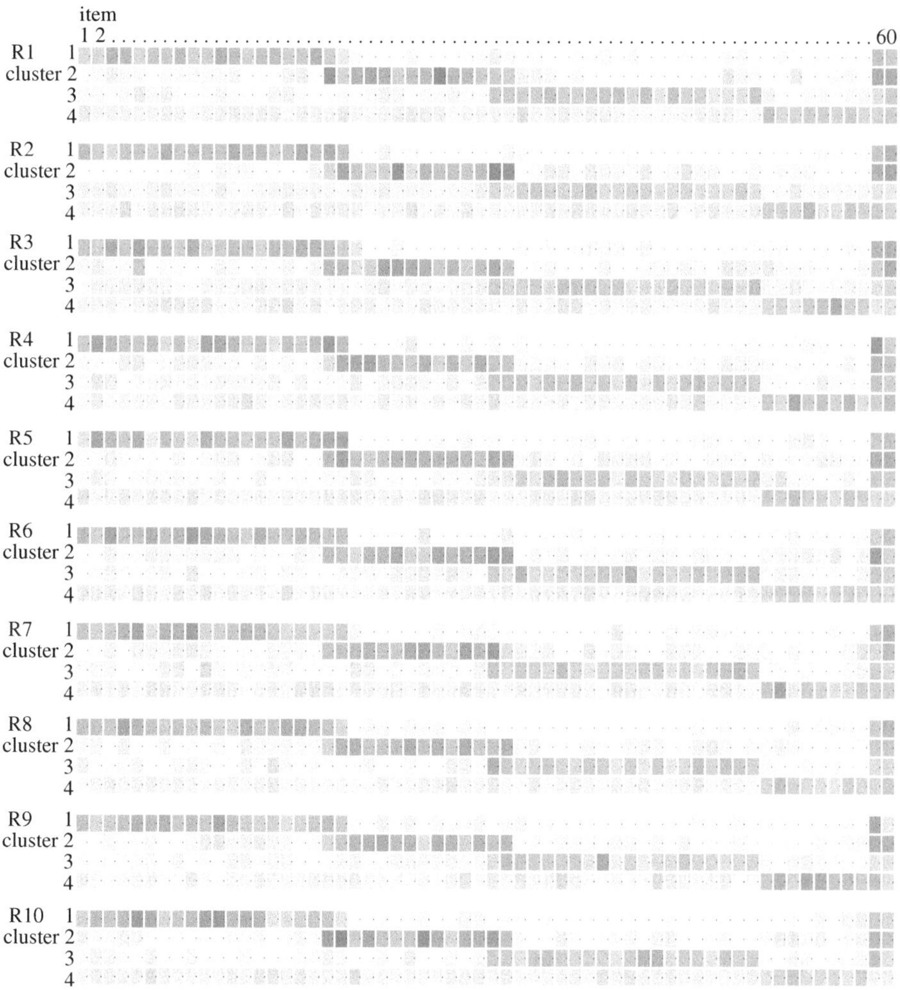

Fig. 4. Comparison of item memberships given by rseFCCM. (Membership degrees are depicted by color gradation.)

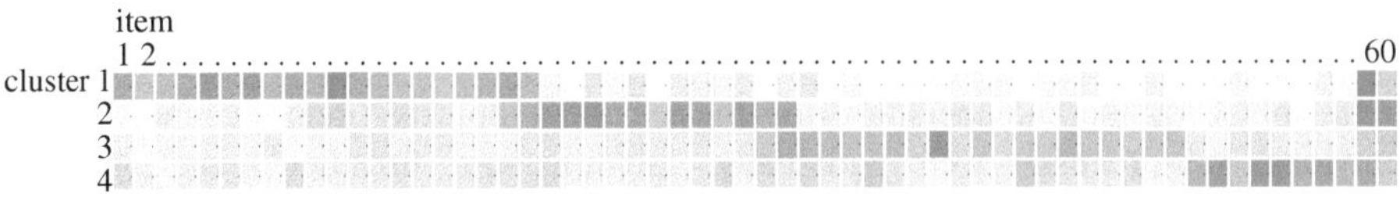

Fig. 5. Item memberships given by spFCCM. (Membership degrees are depicted by color gradation.)

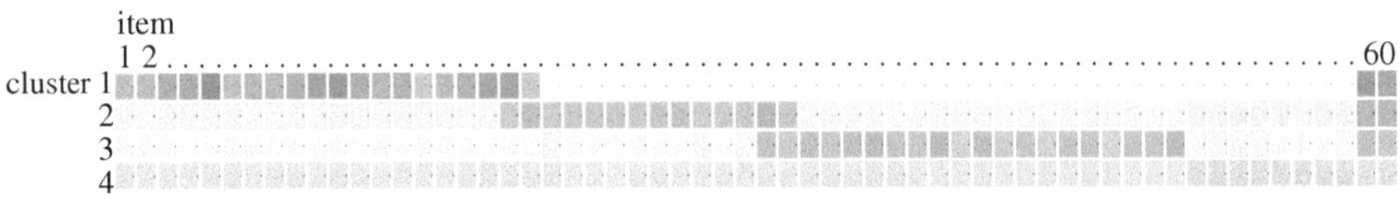

Fig. 6. Item memberships given by oFCCM. (Membership degrees are depicted by color gradation.)

Finally, the oFCCM algorithm was applied. In oFCCM, the conventional FCCM algorithm was first applied to $R_1,\cdots,R_{10}$ in parallel with $C=4$, $\lambda_u=0.001$ and $\lambda_w=100.0$. Note that the derive item memberships in this stage are equivalent to Fig. 4. Then, the 4×10 item membership vectors were gathered into a $(4\times 10)\times 60$ matrix $\tilde{R}$ with their membership weights $\alpha_i^k=\sum_{i=1}^{n_s} u_{ci}^k$. Fig. 5 shows the derives final result.

Here, we should note that the second stage was performed with $C=4$, $\lambda_u=0.00005$ and $\lambda_w=0.5$, which is quite hardened from the first stage. Using the same parameter, the item memberships were very fuzzy and the cluster structure could not be revealed. So, the second stage was performed in a very hard situation but the derived results were still fuzzy. The average correlation coefficient with Fig. 2 was 0.981 and that was very successful result.

As also seen in spFCCM, we can derive a sufficiently reliable result only if the additional memberships are properly hardened. How to harden them should be studied in future work.

5 Conclusion

In this paper, several sampling approaches in FCM clustering of very large data sets were extended to fuzzy co-clustering. Although FCM uses cluster centers as the representative prototypes of clusters and we can identify them with representative object, FCCM does not use such representative prototype. Instead of the cluster centers in FCM, the item memberships are used for characterizing each co-cluster in FCCM and they can be inherited to the remaining sub-processes.

Beside the FCM case, some results implied that additional memberships should be hardened for deriving sufficient results. In future work, the appropriate hardening strategy should be investigated. The applicability to other fuzzy co-clustering models such as Fuzzy CoDoK [10] and SCAD-based model [11] should also be investigated.

Acknowledgements. This work was supported in part by the Ministry of Education, Culture, Sports, Science and Technology, Japan, under Grant-in-Aid for Scientific Research (23500283).

References

1. Havens, T.C., Bezdek, J.C., Leckie, C., Hall, L.O., Palaniswami, M.: Fuzzy *c*-means algorithms for very large data. IEEE Transactions on Fuzzy Systems 20(6), 1130–1146 (2012)
2. Bezdek, J.C.: Pattern Recognition with Fuzzy Objective Function Algorithms. Plenum Press (1981)
3. Pal, N., Bezdek, J.: Complexity reduction for "large image" processing. IEEE Trans. Syst., Man, Cybern. 32(5), 598–611 (2002)
4. Hore, P., Hall, L., Goldgof, D.: Single pass fuzzy c means. In: Proc. IEEE Int. Conf. Fuzzy Syst., pp. 1–7 (2007)
5. Hore, P., Hall, L., Goldgof, D., Gu, Y., Maudsley, A.: A scalable framework for segmenting magentic resonance images. J. Signal Process. Syst. 54(1-3), 183–203 (2009)
6. Oh, C.-H., Honda, K., Ichihashi, H.: Fuzzy clustering for categorical multivariate data. In: Proc. of Joint 9th IFSA World Congress and 20th NAFIPS International Conference, pp. 2154–2159 (2001)
7. Miyamoto, S., Ichihashi, H., Honda, K.: Algorithms for Fuzzy Clustering. Springer (2008)
8. Honda, K., Oh, C.-H., Matsumoto, Y., Notsu, A., Ichihashi, H.: Exclusive partition in FCM-type co-clustering and its application to collaborative filtering. International Journal of Computer Science and Network Security 12(12), 52–58 (2012)
9. Honda, K., Muranishi, M., Notsu, A., Ichihashi, H.: FCM-type cluster validation in fuzzy co-clustering and collaborative filtering applicability. International Journal of Computer Science and Network Security 13(1), 24–29 (2013)
10. Kummamuru, K., Dhawale, A., Krishnapuram, R.: Fuzzy co-clustering of documents and keywords. In: Proc. 2003 IEEE Int'l Conf. Fuzzy Systems, vol. 2, pp. 772–777 (2003)
11. Frigui, H., Nasraoui, O.: Simultaneous categorization of text documents and identification of cluster-dependent keywords. In: Proc. 2002 IEEE Int'l Conf. Fuzzy Systems, vol. 2, pp. 1108–1113 (2002)

Xie-Beni-Type Fuzzy Cluster Validation in Fuzzy Co-clustering of Documents and Keywords

Mai Muranishi, Katsuhiro Honda, and Akira Notsu

Graduate School of Engineering, Osaka Prefecture University,
1-1 Gakuen-cho, Nakaku, Sakai, Osaka 599-8531 Japan
{honda,notsu}@cs.osakafu-u.ac.jp

Abstract. Xie-Beni-type cluster validity indices have been often used for evaluating the quality of Fuzzy c-Means (FCM) cluster partitions because they can validate fuzzy partitions considering the geometrical features of clusters, which suit human feelings in most cases. In Xie-Beni-type indices, cluster compactness and separateness are measured by using intra-cluster deviations and inter-cluster distance (distances among cluster centers). In order to apply Xie-Beni-type indices to co-clustering tasks, the compactness and separateness measures must be modified for handling centroid-less partitions. In this paper, the applicability of a Xie-Beni-type co-cluster validity index to Fuzzy Co-clustering of Documents and Keywords (Fuzzy CoDoK) is investigated.

Keywords: Fuzzy clustering, Co-clustering, Cluster validity.

1 Introduction

Fuzzy c-Means (FCM) [1], [2] has been frequently used in conjunction with various validity measures because FCM can provide various cluster partitions with different settings such as cluster numbers, initial cluster centers and fuzzification weights. In order to find the optimal solution, the best one must be selected considering such validity indices. Partition Coefficient (PC) [1] and Partition Entropy (PE) [3] are used for evaluate the quality of fuzzy partitions, which often prefer *crisp-like* partitions having clear cluster boundaries. Although the indices can be calculated using only fuzzy memberships, they often have monotonically increasing (or decreasing) features. So, the geometric features of fuzzy partitions, which suit human feelings, were employed for measuring the partition quality of prototype-based fuzzy clustering. Xie and Beni [4] considered the compactness and separateness of FCM-type clusters and proposed a cluster validity measure of the ratio of compactness and separateness. We also have several variants of Xie-Beni index [5], [6], [7]. In this paper, the FCM-type cluster validation is further studied in fuzzy co-clustering contexts.

Co-clustering (or Bi-clustering) is the technique for clustering co-occurrence matrices such as document-keyword co-occurrence information in document clustering, user-item purchase history in personalized recommendation problem.

Y.I. Cho et al. (eds.), *Soft Computing in Artificial Intelligence*,
Advances in Intelligent Systems and Computing 270,
DOI: 10.1007/978-3-319-05515-2_4, © Springer International Publishing Switzerland 2014

Fuzzy clustering for categorical multivariate data (FCCM) [8] is an FCM-type co-clustering model, which simultaneously estimates fuzzy partition (fuzzy memberships) of both objects and items by an FCM-like iterative procedure. The FCM clustering criterion was replaced with the degree of aggregation to be maximized without using cluster prototypes. In order to evaluate the quality of FCCM partitions, a Xie-Beni-type cluster index was proposed [9], in which cluster compactness and separateness are measured by considering intra-cluster and inter-cluster aggregations.

In this paper, the Xie-Beni-type co-cluster validity index is applied to another fuzzy co-clustering model of Fuzzy Co-clustering of Documents and Keywords (Fuzzy CoDoK) [10]. Fuzzy CoDoK is a modified algorithm of FCCM, in which membership fuzzification is achieved by the quadric regularization approach [2] and is available with large data sets.

In Section 2, we give a brief review on the conventional FCM-type cluster validity indices and co-clustering model. In Section 3, we introduce a Xie-Beni-type validity index, which can be used in fuzzy co-clustering context. In Section 4, we present several experimental results to demonstrate the characteristic features of the proposed index by applying to FCCM and Fuzzy CoDoK. In Section 5, we summarize our conclusions.

2 FCM Cluster Validation and Fuzzy Co-clustering

2.1 FCM and Cluster Validation

Let $\boldsymbol{x}_i$, $i = 1,\ldots,n$ be the m dimensional observation of n objects. FCM partitions the objects into C fuzzy clusters by using prototypical centroid $\boldsymbol{b}_c$, $c = 1,\ldots,C$. Object assignment is represented by fuzzy membership u_{ci} i.e., the degree of belongingness of object i to cluster c. The clustering criterion is defined by the distance between objects and centroids in the multi-dimensional space as:

$$L_{fcm} = \sum_{c=1}^{C} \sum_{i=1}^{n} u_{ci}^{\theta} \|\boldsymbol{x}_i - \boldsymbol{b}_c\|^2, \tag{1}$$

where θ (θ>1) is an exponential weight for membership fuzzification [1] and the model is reduced to the crisp k-Means [11] if θ=1. The sum of u_{ci} with respect to c is usually constrained to be 1.

First, the FCM partition quality was measured by Partition Coefficient (PC) [1] and Partition Entropy (PE) [3], which measure the crispness of fuzzy memberships as:

$$PC = \frac{1}{n} \sum_{c=1}^{C} \sum_{i=1}^{n} u_{ci}^2, \tag{2}$$

$$PE = -\frac{1}{n} \sum_{c=1}^{C} \sum_{i=1}^{n} [u_{ci} \log u_{ci}] \tag{3}$$

PC (or PE) becomes smaller (or larger) as fuzzy memberships become very fuzzy. So, the optimal partition can be the one having large PC (or small PE).

Although the indices can directly measure the partition quality, the selected partition sometimes does not suit our human sense because they lack geometric feelings. Considering the geometric features in the multi-dimensional space, Xie and Beni [4] proposed a measure for evaluating both the compactness and separateness of fuzzy clusters as:

$$V_{XB} = \frac{\sum_{c=1}^{C}\sum_{i=1}^{n} u_{ci}^{2} \|\boldsymbol{x}_i - \boldsymbol{b}_c\|^2}{n\left(\min_{k,l} \|\boldsymbol{b}_k - \boldsymbol{b}_l\|^2\right)}, \tag{4}$$

where the numerator is the FCM objective function with θ=2 and measures the compactness of clusters while the denominator measures the separateness of clusters. The optimal compact/separate cluster can be found by minimizing the Xie-Beni index.

2.2 FCM-Type Co-clustering

2.2.1 Fuzzy Clustering for Categorical Multivariate Data (FCCM)

In order to extract co-cluster structures from co-occurrence information among objects and items, co-clustering simultaneously partitions both objects and items into bi-clusters composed of familiar objects and items. Let $R = \{r_{ij}\}$ is a similarity (co-occurrence) matrix, where r_{ij} is the similarity degree ($0 \le r_{ij} \le 1$) among object i, (i=1,...,n) and item j, (j=1,...,m).

In FCCM, the objective function is defined by considering the aggregation degree of each cluster as:

$$L_{fccm} = \sum_{c-1}^{C}\sum_{i=1}^{n}\sum_{j=1}^{m} u_{ci} w_{cj} r_{ij} - \lambda_u \sum_{c=1}^{C}\sum_{i=1}^{n} u_{ci} \log u_{ci} - \lambda_w \sum_{c=1}^{C}\sum_{j=1}^{m} w_{cj} \log w_{cj}. \tag{5}$$

u_{ci} and w_{cj} are the fuzzy memberships of object i and item j to cluster c, respectively. Object memberships u_{ci} represent the cluster assignment of objects and are given by considering the FCM-like exclusive condition where the sum of u_{ci} with respect to c is forced to be 1. On the other hand, avoiding a trivial solution, item memberships w_{cj} are estimated by considering a intra-cluster condition where the sum of w_{cj} with respect to j is forced to be 1 in each cluster. So, the item memberships represent relative responsibility in each cluster and each item can be shared (rejected) by multiple (all) clusters. Fuzzification of memberships was achieved by the entropy-based regularization approach [2] and the entropy terms play a similar role with the weighting exponent in FCM. λ_u and λ_w tune the degree of fuzziness of memberships and larger values give fuzzier partitions. The clustering algorithm is an FCM-type iterative optimization procedure of u_{ci} and w_{cj}. The updating rules for u_{ci} and w_{cj} are given as:

$$u_{ci} = \frac{\exp\left(\lambda_u^{-1}\sum_{j=1}^{m} w_{cj} r_{ij}\right)}{\sum_{l=1}^{C}\exp\left(\lambda_u^{-1}\sum_{j=1}^{m} w_{lj} r_{il}\right)}, \tag{6}$$

$$w_{cj} = \frac{\exp\left(\lambda_w^{-1}\sum_{i=1}^{n} u_{ci} r_{ij}\right)}{\sum_{l=1}^{n}\exp\left(\lambda_u^{-1}\sum_{j=1}^{n} u_{ci} r_{il}\right)}. \tag{7}$$

2.2.2 Fuzzy Co-clustering of Documents and Keywords(Fuzzy CoDoK)

For large values of n and m, the numerators and denominators of FCCM updating function of (6) and (7) can become extremely large. This can lead to numerical instabilities due to overflows. Then, the objective function was modified by introducing the fuzzification approach with the quadric terms [2] instead of entropy terms as the following Fuzzy CoDoK:

$$L_{codok} = \sum_{c=1}^{C}\sum_{i=1}^{n}\sum_{j=1}^{m} u_{ci} w_{cj} r_{ij} - T_u \sum_{c=1}^{C}\sum_{i=1}^{n} u_{ci}^2 - T_w \sum_{c=1}^{C}\sum_{j=1}^{m} w_{cj}^2. \tag{8}$$

T_u and T_w are the fuzzification weights, which can be identified with λ_u and λ_w in FCCM. The updating rules for the two memberships are derived by considering the optimality conditions of the objective function as:

$$u_{ci} = \frac{1}{C} + \frac{1}{2T_u}\left[\sum_{j=1}^{m} w_{cj} r_{ij} - \frac{1}{C}\sum_{l=1}^{C}\sum_{j=1}^{m} w_{lj} r_{ij}\right], \tag{9}$$

$$w_{cj} = \frac{1}{m} + \frac{1}{2T_v}\left[\sum_{i=1}^{n} u_{ci} r_{ij} - \frac{1}{m}\sum_{l=1}^{m}\sum_{i=1}^{n} w_{lj} r_{ij}\right]. \tag{10}$$

Here, u_{ci} and w_{cj} given by the above formulations can be negative. In a simple strategy, negative memberships are replaced with *zero-memberships* and the remaining *positive* memberships are renormalized so as to sum to one.

3 Cluster Validation in FCM-Type Co-clustering

3.1 Applicability of Partition Coefficient and Partition Entropy

In order to evaluate the partition quality of fuzzy partitions, PC and PE can be applied to 2 different types of fuzzy memberships. In the conventional PC and PE of Eqs.(2) and (3), the measure was normalized by n because the sum-to-one condition with respect to c brings $\sum_{c=1}^{C}\sum_{i=1}^{n} u_{ci} = n$. On the other hand, when the sum-to-one condition with respect to j is used for item memberships w_{cj}, it brings $\sum_{c=1}^{C}\sum_{j=1}^{m} w_{cj} = C$. Then, they should be slightly modified for measuring the quality of co-cluster partitions as:

$$PC_u = \frac{1}{n}\sum_{c=1}^{C}\sum_{i=1}^{n} u_{ci}^2, \qquad PC_w = \frac{1}{C}\sum_{c=1}^{C}\sum_{j=1}^{m} w_{cj}^2, \tag{11}$$

$$PE_u = -\frac{1}{n}\sum_{c=1}^{C}\sum_{i=1}^{n}\left[u_{ci}\log u_{ci}\right], \quad PE_u = -\frac{1}{C}\sum_{c=1}^{C}\sum_{j=1}^{m}\left[w_{cj}\log w_{cj}\right] \tag{12}$$

Then, the total quality of dual partition is summarized such combinations as:

$$PC_{co} = PC_u \times PC_w, \tag{13}$$

$$PE_{co} = PE_u \times PE_w. \tag{14}$$

3.2 A New Compactness/Separateness Index

Honda et al. [12] proposed a Xie-Beni-type validity measure for co-cluster validation. In the measure, the intra-cluster compactness is measured by

$$compactness = \frac{1}{C}\sum_{c=1}^{C}\sum_{i=1}^{n}\sum_{j=1}^{m} u_{ci}w_{cj}\left(d_{ij} - \left(1 - d_{ij}\right)\right), \tag{15}$$

where d_{ij} contributes to evaluation of the aggregation degree while (1- d_{ij}) reject null-pair of user i and item j. So, a large value implies a densely and tightly connected cluster partition. The larger the value of Eq.(15), the more compact the partition is.

Next, the inter-cluster similarity is measured by the connection degrees among users and items belonging to different clusters. An average measure for inter-cluster separateness of $C(C$-1) cluster pairs is measured by

$$separateness = \frac{1}{C\left(C-1\right)}\sum_{k=1}^{C}\sum_{l\neq k}\sum_{i=1}^{n}\sum_{j=1}^{m} u_{ki}w_{lj}d_{ij}. \tag{16}$$

The smaller the value of Eq.(16) is, the more separate the partition is. Combining the above compactness and separateness measures, a Xie-Beni-like validation measure for fuzzy co-cluster partition V_{XBco} is defined as:

$$V_{XB_{co}} = \frac{compactness}{separateness} = \frac{(C-1)\sum_{c=1}^{C}\sum_{i=1}^{n}\sum_{j=1}^{m} u_{ci} w_{cj}\left(2d_{ij}-1\right)}{\sum_{k=1}^{C}\sum_{l\neq k}\sum_{i=1}^{n}\sum_{j=1}^{m} u_{ki} w_{lj} d_{ij}} \tag{17}$$

The larger the value of V_{XBco} is, the more compact and separate the fuzzy co-cluster partition is. So, we can find the optimal co-cluster partition by searching for the largest V_{XBco}.

4 Numerical Experiment

4.1 Artificial Data Sets

First, the proposed validity measure is applied to an artificial data set. The co-occurrence matrices shown in Fig.1 includes roughly 4 square-like co-clusters having different sizes or shapes, in which black squares means r_{ij}=1 and otherwise r_{ij}=0. Some items are shared by multiple clusters. The FCCM algorithm and Fuzzy CoDoK algorithm were applied to the data set with various cluster numbers C={2,3,...,10}, and the fuzzification weights were set as λ_u=0.05 and λ_w=5.0. Table 1 compares the values of several validity indices and the selected cluster number whose index values are given by bold.

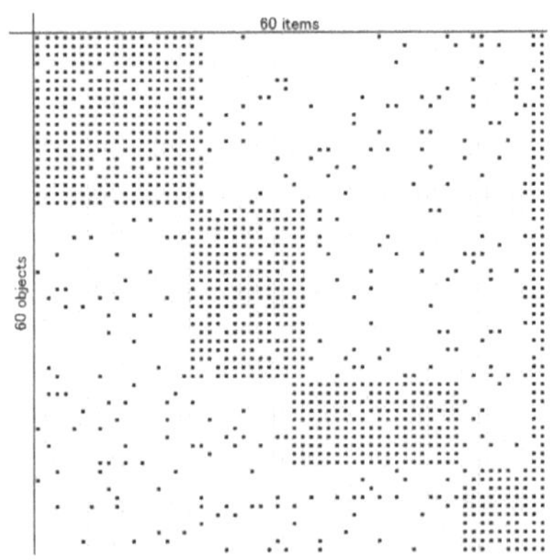

Fig. 1. Artificial Data Set

In order to confirm the applicability of V_{XBco}, the values are compared not only with each of PC and PE for objects and items but also with the products of them in the Table 1. Using both fuzzy co-clustering models, PC and PE could not find the optimal cluster number of C=4 but selected C=2 as the best one because of the influence of the shared items. The proposed index could select the optimal cluster number of C=4. The reason why the proposed V_{XBco} works well in such sharing situations despite of instability of PC and PE is that the compactness and separateness measures fairly calculate the partition quality even if fuzzy memberships for the shared items become more ambiguous than dual exclusive situations.

Table 1. Comparison of clustering results with artificial data set

	FCCM							Fuzzy CoDok						
C	PC_u	PC_w	PC_{co}	PE_u	PE_w	PE_{co}	V_{XBco}	PC_u	PC_w	PC_{co}	PE_u	PE_w	PE_{co}	V_{XBco}
2	**0.993**	**0.113**	**0.112**	**0.017**	**2.839**	**0.050**	1.81	**1.0**	**0.058**	**0.058**	**0**	2.985	**0**	1.87
3	0.980	0.051	0.050	0.041	3.367	0.139	1.82	0.980	0.050	0.049	0.029	3.049	0.090	2.53
4	0.988	0.038	0.037	0.028	3.532	0.100	**1.98**	0.992	0.063	0.063	0.016	2.824	0.044	**3.45**
5	0.984	0.034	0.033	0.039	3.645	0.143	1.85	0.826	0.063	0.052	0.247	**2.810**	0.694	2.74
6	0.892	0.030	0.026	0.227	3.752	0.851	1.55	0.723	0.058	0.042	0.396	2.909	1.151	2.74
7	0.884	0.028	0.024	0.259	3.802	0.985	1.51	0.576	0.059	0.034	0.664	2.886	1.917	2.32
8	0.878	0.026	0.023	0.287	3.839	1.101	1.49	0.514	0.059	0.030	0.815	2.870	2.340	2.18
9	0.873	0.025	0.022	0.311	3.868	1.205	1.47	0.482	0.060	0.029	0.925	2.857	2.642	2.08
10	0.869	0.024	0.021	0.334	3.891	1.299	1.45	0.461	0.060	0.028	1.012	2.848	2.881	2.01

This result implies that the cluster validation index V_{XBco} is still useful for Fuzzy CoDoK.

5 Document Clustering

Second, the 2 fuzzy co-clustering models were applied to a document clustering task. The data set used in [13] was constructed from a Japanese novel ``Kokoro" written by Soseki Natsume, which can be downloaded from Aozora Bunko (http://www.aozora.gr.jp). The novel is composed of 3 chapters, each of which include 36, 18, 56 sections, respectively. In this experiment, the sections were given as objects (n=110) and the co-occurrence frequencies with 83 most frequently used substantives and verbs were used for constructing a co-occurrence matrix (m=83), whose elements are their normalized tf-idf weights [14] so that each of column elements has same average and variance and is bounded in the interval [0,1].

The FCCM algorithm and Fuzzy CoDoK algorithm were applied with C={2,..., 7}. The fuzzification weights of the FCCM are λ_u=0.01 and λ_w=1.0. The fuzzification weights of the Fuzzy CoDoK are λ_u=0.00001 and λ_w=1.0. The sections and keywords were partitioned into clusters without using the chapter information. Table 2 compares the values of indices. In the same manner with the previous subsection, PC-type and PE-type validity indices did not derive stable results, in which both u_{ci} and w_{cj} are validated in a same cluster number, and they always selected C<5. On the other hand, in FCCM, V_{XBco} selected C=5, and in the Fuzzy CoDoK, it setected C=6.

At a glance, the results may be seen to be doubtful because the actual chapter number is 3, and the results are different in FCCM and Fuzzy CoDoK.

In order to validate the optimality of C=5 and C=6, Fig.2 depicts the cluster memberships of each section (object) and Table 3 compares the keywords (items) having the top 5 memberships in each cluster. Fig.2 indicates that the 5 clusters quite fit the story and the 3rd chapter was shown to be composed of three sub-stories. In the case of Fuzzy CoDoK, the 1st chapter composed of 2 sub-stories, and the 3rd chapter is composed of 3 sub-stories. The rough image of each cluster can be summarized from the keyword shown in Table 3. Cluster 1-1, and Cluster 1-2 correspond to chapter 1, whose chapter title is *Sensei and I*, and the selected keywords imply a story on the problem related to *sensei* (master) and the protagonist. Cluster 2 corresponds to chapter 2, whose chapter title is *My Parents and I*, and the selected keywords imply a story on his parents. Cluster 3 corresponds to the first and the last part of chapter 3, whose chapter title is *Sensei and His Testament*, and the selected keywords imply a story on his family. (Indeed, the first part of chapter 3 is related to his family story and it is revisited in the last part.) Cluster 4 corresponds to the second part of chapter 3, which includes a story on a lady. (In the part, the protagonist falls in love with a lady.) Cluster 5 corresponds to the main part of chapter 4, which includes a story on *K* and the lady. (In the part, he had a trouble with *K* on the lady.) These summaries are quite useful for intuitively understand the rough story of the novel. In this way, FCCM or Fuzzy CoDoK have ability for summarizing the contents of documents by selecting the optimal cluster number based on the proposed validity index.

Table 2. In the document clustering, the left result is FCCM and the right one is Fuzzy CoDoK

	FCCM							Fuzzy CoDok						
C	PC_u	PC_w	PC_{co}	PE_u	PE_w	PE_{co}	V_{XBco}	PC_u	PC_w	PC_{co}	PE_u	PE_w	PE_{co}	V_{XBco}
2	**0.945**	0.061	0.058	**0.096**	3.573	**0.344**	-2.12	**1.0**	0.042	0.042	**0**	3.377	**0**	-2.12
3	0.929	0.094	0.087	0.130	**3.366**	0.439	--1.89	0.966	0.048	0.046	0.053	3.256	0.172	-1.91
4	0.943	0.101	**0.095**	0.115	3.428	0.394	-1.75	0.906	0.053	**0.048**	0.143	3.183	0.455	-1.79
5	0.901	**0.804**	0.076	0.196	3.621	0.710	**-1.71**	0.806	**0.057**	0.046	0.300	3.134	0.940	-1.72
6	0.806	0.071	0.057	0.381	3.788	1.443	-1.71	0.704	0.057	0.040	0.470	**3.114**	1.463	**-1.68**
7	0.592	0.044	0.026	0.871	4.036	3.515	-1.81	0.689	0.045	0.031	0.577	3.284	1.894	-1.74

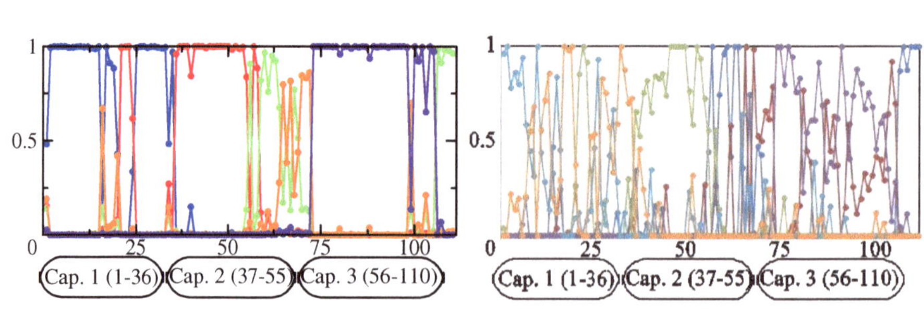

Fig. 2. Memberships of Sections

Table 3. Top 5 Keywords in Each Cluster (document clustering)

	FCCM					Fuzzy CoDoK				
C	1st	2nd	3rd	4th	5th	1st	2nd	3rd	4th	5th
1-1	**sensei (master)**	**kore (this)**	**mieru (look)**	**iku (go)**	**mondai (problem)**	**sensei (master)**	**hito (person)**	**mieru (look)**	**iku (go)**	**doko (where)**
1-2						**sensei (master)**	**kiku (hear)**	**okusan (mistless)**	**mondai (problem)**	**kore (this)**
2	chichi (father)	haha (mother)	tegami (letter)	byouki (illness)	kaku (write)	chichi (father)	haha (mother)	tegami (letter)	dasu (send)	kaku (write)
3	sai (wife)	oji (uncle)	watashi (I)	kokoro (heart)	miru (watch)	sai (wife)	oji (uncle)	watashi (I)	anata (you)	jibun (I)
4	ojosan (lady)	otoko (male)	okusan (mistless)	hanashi (talk)	miru (watch)	ojosan (lady)	heya (room)	suwaru (sit)	miru (watch)	watashi (I)
5	K (a name)	kare (he)	mukau (face)	ojosan (lady)	heya (room)	kare (he)	K (a name)	mukau (face)	toori (street)	kotae (answer)

6 Conclusions

In this paper, a fuzzy co-cluster validity measure was using with both FCCM and Fuzzy CoDoK and its applicability was compared. Several experimental results imply that the validity measure, which was first proposed for FCCM, was also applicable to Fuzzy CoDoK although several classical measures such as PC and PE are not available. Possible future work includes a further study on the applicability to other FCM-based co-clustering models such as SCAD-based ones [15].

Acknowledgements. This work was supported in part by the Ministry of Education, Culture, Sports, Science and Technology, Japan, under Grant-in-Aid for Scientific Research (23500283, 23500284).

References

[1] Bezdek, J.C.: Pattern Recognition with Fuzzy Objective Function Algorithms. Plenum Press (1981)

[2] Miyamoto, S., Ichihashi, H., Honda, K.: Algorithms for Fuzzy Clustering: Methods in c-Means Clustering with Applications. STUDFUZZ, vol. 229. Springer, Heidelberg (2008)

[3] Bezdek, J.C.: Cluster validity with fuzzy sets. J. Cybernet 3, 58–73 (1974)

[4] Xie, X.L., Beni, G.: A validity measure for fuzzy clustering. IEEE Trans. Pattern Anal. Mach. Intell. 13, 841–847 (1991)

[5] Fukuyama, Y., Sugeno, M.: A new method of choosing the number of clusters for the fuzzy c-means method. In: Proc. Fifth Fuzzy Systems Symp., pp. 247–250 (1989)

[6] Kwon, S.H.: Cluster validity index for fuzzy clustering. Electron. Lett. 34(22), 2176–2177 (1998)

[7] Rezaee, M.R., Lelieveldt, B.P.F., Reiber, J.H.C.: A new cluster validity index for the fuzzy c-mean. Pattern Recognition Lett. 19, 237–246 (1998)
[8] Oh, C.-H., Honda, K., Ichihashi, H.: Fuzzy clustering for categorical multivariate data. In: Proc. of Joint 9th IFSA World Congress and 20th NAFIPS International Conference, pp. 2154–2159 (2001)
[9] Muranishi, M., Honda, K., Notsu, A., Ichihashi, H.: A cluster Validity Index for FCM-type Co-clustering. In: Proc. 2013 IEEE Int'l Conf. Fuzzy Systems, paper #1161, pp. 1–6 (2013)
[10] Kummamuru, K., Dhawale, A., Krishnapuram, R.: Fuzzy Co-clustering of Documents and Keywords. In: Proc. 2003 IEEE Int'l Conf. Fuzzy Systems, vol. 2, pp. 772–777 (2003)
[11] MacQueen, J.B.: Some methods of classification and analysis of multivariate observations. In: Proc. of 5th Berkeley Symposium on Math. Stat. and Prob., pp. 281–297 (1967)
[12] Honda, K., Muranishi, M., Notsu, A., Ichihashi, H.: FCM-type Cluster Validation in Fuzzy Co-clustering and Collaborative Filtering Applicability. International Journal of Computer Science and Network Security 13(1), 24–29 (2013)
[13] Honda, K., Notsu, A., Ichihashi, H.: Fuzzy PCA-guided robust k-means clustering. IEEE Transactions on Fuzzy Systems 18(1), 67–79 (2010)
[14] Salton, G., Buckley, C.: Term-weighting approaches in automatic text retrieval. Information Processing and Management 24(5), 513–523 (1988)
[15] Frigui, H., Nasraoui, O.: Simultaneous categorization of text documents and identification of cluster-dependent keywords. In: Proc. 2002 IEEE Int'l Conf. Fuzzy Systems, vol. 2, pp. 1108–1113 (2002)

Fuzzy c-Regression Models Based on Optimal Scaling of Categorical Observation with Tolerance

Shunsuke Iwata, Katsuhiro Honda, and Akira Notsu

Dept. of Computer Science and Intelligent Science, Osaka Prefecture University
1-1 Gakuen-cho, Nakaku, Sakai, Osaka 599-8531, Japan
{honda,notsu}@cs.osakafu-u.ac.jp

Abstract. Switching regression is a powerful tool for revealing intrinsic non-linear dependencies among exploratory variables and objective variables. In this paper, Fuzzy c-Regression Models (FCRM), which is an FCM-type switching regression model, is modified so that it can handle uncertain categorical observations. In data mining applications, we often deal with databases consisting of mixed measurement levels. The alternating least squares method is a technique for mixed measurement situations, in which categorical observations are quantified such that they suit the current model by optimal scaling, and has been applied to FCM-type fuzzy clustering in order to characterize each cluster considering mutual relation among categories. While optimal scaling has been used for handling categorical observations, categorical observations often have ambiguity of natural language categories. In this paper, a modified FCM-type switching regression is performed by considering tolerance of quantified category observations in conjunction with optimal scaling.

Keywords: fuzzy clustering, switching regression, optimal scaling, ambiguity.

1 Introduction

Switching regression [1], [2], which estimates multiple local regression functions, is a powerful tool for revealing intrinsic non-linear dependencies among exploratory variables and objective variables. Fuzzy c-Regression Models (FCRM) [3] is an FCM-type switching regression algorithm, which iteratively estimate fuzzy memberships of objects and local regression functions in an FCM-like procedure [4]. In FCRM, each regression function is identified with the representative prototype of a cluster and the regression errors are also used as the clustering criterion.

In data mining applications, we often face to mixed measurement situations where we need to characterize intrinsic structures of multivariate data considering both of numerical and categorical observations. The alternating least squares (ALS) method [5], [6], [7] is a technique for performing statistical analysis of data measured at a variety of levels. The ALS-based iterative algorithms alternate two phases, each of which is a least squares procedure. One is the data transformation phase for obtaining

Y.I. Cho et al. (eds.), *Soft Computing in Artificial Intelligence*,
Advances in Intelligent Systems and Computing 270,
DOI: 10.1007/978-3-319-05515-2_5, © Springer International Publishing Switzerland 2014

least squares estimates for data quantification based on optimal scaling while model parameters are held constant. The other is the (conventional) model estimation phase for obtaining least squares estimates for model parameters while the data transformations are held constant. However, in the case of that each category is characterized by natural language, their quantified scores may have ambiguity.

In this paper, a new FCM-type switching regression algorithm is proposed, which can handle categorical observations considering the tolerance of their quantified scores. By introducing the concept of tolerance, FCRM is performed in a robust way. The characteristic features of the proposed method are demonstrated in a numerical experiment.

2 FCM-Type Switching Regression and Handling Uncertain Data

2.1 Fuzzy *c*-Regression Models

FCM [4], [8] is a fuzzy variant of *k*-Means clustering [9], which is the prototype-based non-hierarchical clustering method. Let x_i, $i=1,\dots,n$ be m-dimensional observation on n samples and $\boldsymbol{b}_c = 1,\dots,C$ be prototypical centroids of C clusters. The clustering criterion to be minimized is the distance $d_{ci}=||\boldsymbol{x}_i-\boldsymbol{b}_c||$ and the objective function is defined as follows:

$$L_{fcm} = \sum_{c=1}^{C}\sum_{i=1}^{n} u_{ci}^{\theta} d_{ci} \tag{1}$$

where u_{ci} is the membership degree of sample i to cluster c and is estimated under the sum-to-one condition of $\sum_{c=1}^{C} u_{ci} = 1$ for all i. The *k*-Means-type algorithm updates fuzzy memberships u_{ci} and prototypes $\boldsymbol{b}_c$ until convergent.

FCRM [3] is the FCM-type switching regression algorithm, in which the regression errors to be minimized is also used for clustering criteria. Let y_i be the objective variable corresponding to exploratory observation $\boldsymbol{x}_i$ and the goal be to predict y_i from $\boldsymbol{x}_i$. In this paper, local linear regression models are considered. In FCRM, the n samples are partitioned into c clusters in conjunction with estimation of the following local regression models in each cluster:

$$y_c \approx a_{c0} + a_{c1}x_1 + \dots + a_{cm}x_m \tag{2}$$

where $\boldsymbol{a_c} = (a_{c0}, a_{c1}, \dots, a_{cm})^T$ is the regression coefficient vector to be estimated in cluster c. The FCM-type iterative algorithm is performed by replacing clustering criterion d_{ci} with regression error $e_{ci} = |y_{ci} - a_{c0} - a_{c1}x_{i1} - \dots - a_{cm}x_{im}|$. Fuzzy membership u_{ci} is updated by

$$u_{ci} = \begin{cases} \left[\sum_{l=1}^{C} \left(\frac{\|e_{ci}\|^2}{\|e_{li}\|^2} \right)^{\frac{1}{\theta-1}} \right]^{-1} & ; I_i = 0 \\ \frac{1}{|I_i|} & ; I_i \neq 0, c \in I_i \\ 0 & ; I_i \neq 0, c \notin I_i \end{cases} \tag{3}$$

where I_i is the set of indices such that e_{ci} is zero, i.e., $I_i = \{c | e_{ci} = 0\}$. Regression coefficient vector $\boldsymbol{a}_c$ is updated by

$$\boldsymbol{a}_c = \left[\tilde{X}^T U_c^\theta \tilde{X} \right]^{-1} \tilde{X}^T U_c^\theta \boldsymbol{y}, \tag{4}$$

where $\boldsymbol{y}$ is $\boldsymbol{y} = (y_1, \dots, y_n)^T$ and $\tilde{X}$ is the $(n \times (m+1))$ matrix whose i-th row is $\tilde{x}_i^T = (1, x_i^T)$. $\boldsymbol{U}_c$ is the $(n \times n)$ diagonal matrix whose i-th diagonal element is u_{ci}.

2.2 FCM Clustering of Uncertain Data

When observed data have uncertainty, prototype estimation in FCM should be performed in a robust way. Introducing the concept of tolerance [10], [11], the objective function of FCM was modified, in which ambiguous observation can be adjusted within a possible tolerance range. Assume that an observation x_i has a certain ambiguity and may be adjusted as $\boldsymbol{x}_i \rightarrow \boldsymbol{x}_i + \boldsymbol{\xi}_i$. The constraint on $\boldsymbol{\xi}_i$ can be given in various ways such as possible range [10] or penalty [11]. If we use penalty for minimizing the norm of $\boldsymbol{\xi}_i = (\xi_{i1}, \dots \xi_{ij})$, the FCM objective function was re-defined as follows:

$$L_{fcmt} = \sum_{c=1}^{C} \sum_{i=1}^{n} u_{ci}^\theta \|\boldsymbol{x}_i + \xi_i - \boldsymbol{b}_c\|^2 + \gamma \sum_{i=1}^{n} \|\xi_i\|^2, \tag{5}$$

where γ is the penalty weight. By optimally adjusting the ambiguous observations, we can estimate robust cluster centers.

3 Fuzzy c-Regression Models of Mixed Databases Considering Ambiguity of Category Observations

3.1 Optimal Scaling of Categorical Variables in ALS Method

The j-th numerical observation of the i-th sample is represented as x_{ij} while the j-th categorical variable with K_j categories is given as a K_j dimensional indicator vector $\boldsymbol{g}_{ij}$ whose element g_{ijk} is defined as follows:

$$g_{ijk} = \begin{cases} 1; & \text{if sample } i \text{ belongs to category } k. \\ 0; & \text{otherwise.} \end{cases}$$

In the alternating least squares (ALS) method [5], [6], [7], we quantify categorical observation $\boldsymbol{g}_{ij}$ into numerical score $x_{ij} = \boldsymbol{g}_{ij}^T \boldsymbol{q}_j$ by estimating the optimal quantification vector $\boldsymbol{q}_j$ that is most responsible for representing the current model. After data transformation (optimal scaling), we have only numerical observations and can apply conventional model estimation algorithms for updating the current model. Based on the iterative optimization scheme that alternates the two phases, we can derive optimal solutions, e.g., in the MORALS algorithm [5], the data transformation phase estimates least squares estimates for quantification vectors while the model estimation phase obtains least squares estimates for a partial regression coefficient matrix.

Honda *et al.* [12] applied the optimal scaling approach to FCM clustering and combined the step of optimal scaling of categorical values with estimation of fuzzy memberships of objects and cluster prototypes. Optimal quantification scores are estimated as arithmetic weighted mean of coordinate values of cluster centroids.

3.2 Fuzzy c-Regression Models Based on Optimal Scaling

Assume that we have n samples $(\boldsymbol{x}_i, y_i)$, $i=1,\ldots,n$ composed of m-dimension explanatory variables and one objective variable and the first m q explanatory variables are numerical ones and the remaining q variables are categorical ones. As to the j-th categorical variable with K_j categories, the optimal quantification vector $\boldsymbol{q}_j$ is estimated for each categorical variable. Then, the following criterion was constructed for performing FCM-type switching regression with mixed databases [13]:

$$e_{ci} = \left(y_i - b_{c0} - \sum_{j=1}^{m-q} b_{cj} x_{ij} - \sum_{j=m-q+1}^{m} b_{cj} \boldsymbol{g}_{ij}^T \boldsymbol{q}_j \right)^2 \tag{6}$$

where b_{c0} is the constant term. In this algorithm, the category score vector is calculated one by one. To derive the optimal $\boldsymbol{q}_j$ with fixed $\boldsymbol{q}_k (k \neq j)$, the linear regression model is decomposed as follows:

$$\boldsymbol{y} = \boldsymbol{Z}^* \boldsymbol{b}_c + \boldsymbol{G}_j \boldsymbol{q}_j b_{cj} + \boldsymbol{e} \tag{7}$$

where $\boldsymbol{y}$ is $\boldsymbol{y} = (y_1, \ldots, y_n)^T$. $\boldsymbol{Z}^*$ is the $(n \times m)$ data matrix consisting of constant term, $(m - q)$ numerical variables and $(q - 1)$ quantified scores $x_{il} = \boldsymbol{g}_{il}^T \boldsymbol{q}_l$, $l = m-q+1, \ldots, j-1, j+1, \ldots, m$. $\boldsymbol{b}_c^*$ is the corresponding m-dimensional vectors. b_{cj} is the j-th element of $\boldsymbol{a}_c$. $\boldsymbol{G}_j$ is the $(n \times K_j)$ matrix and $\boldsymbol{G}_j = (\boldsymbol{g}_{1j}, \ldots, \boldsymbol{g}_{nj})^T$. Then, the optimal quantification vector $\boldsymbol{q}_j$ is calculated as:

$$\boldsymbol{q}_j = \left(\boldsymbol{G}_j^T \left(\sum_{c=1}^{C} |b_{cj}|^2 \boldsymbol{U}_c^{\theta} \right) \boldsymbol{G}_j \right)^{-1} \times \left(\sum_{c=1}^{C} \boldsymbol{G}_j^T \boldsymbol{U}_c^{\theta} \left(\boldsymbol{y} - \boldsymbol{Z}^* \boldsymbol{b}_c^* \right) b_{cj} \right) \tag{8}$$

To avoid trivial solutions, the mean of j-th explanatory variables of quantified scores $x_{ij} = \boldsymbol{g}_{ij}^T \boldsymbol{q}_j$ is constrained to be 0 and the variance is forced to be σ^2. The priority weight of categorical variables can be adjusted by the variance parameter σ^2.

3.3 Robust Fuzzy c-Regression Models Considering Tolerance of Ambiguity

In this paper, the concept of tolerance is introduced into FCRM for handling the ambiguity of categorical variables. Categorical observations often have ambiguity of natural language categories. In such a case, the category scores should be adjusted by considering tolerance of quantified category observations in conjunction with optimal scaling. In the rest of this section, robust FCRM is proposed with the concept of tolerance of categorical observation.

Assume that exploratory variable j has the ambiguity of natural language categories and the observation for object i can be adjusted as $x_{ij} \rightarrow x_{ij} + \xi_{ij}$. Considering the adjustment, the objective function of FCRM with local linear regression models is re-defined as follows:

$$L_{fcrmt} = \sum_{c=1}^{C} \sum_{i=1}^{n} u_{ci}^{\theta} \left| y_i - a_{a0} - a_{ci} x_{i1} - \ldots - a_{cj}(x_{ij} + \xi_{ij}) - \quad - a_{cm} x_{im} \right|^2 + \gamma \left\| \xi_j \right\|^2$$
$$s.t \quad \sum_{c=1}^{C} u_{ci} = 1, \ \forall i, j \tag{9}$$

The penalty term is used for constraining the norm of tolerance vector $\boldsymbol{\xi}_j$. By minimizing the squared norm of $\boldsymbol{\xi}_j$, tolerance value ξ_{ij} has a large value only for objects with uncertainty and it is avoided to illegally distort the regression models. The updating rule for tolerances ξ_{ij} is given from $\partial L_{fcrmt} / \partial \xi_{ij} = 0$.

$$\frac{\partial L_{fcrmt}}{\partial \xi_{ij}} = -2 \sum_{c=1}^{C} u_{ci}^{\theta} a_{cj} \left(y_i - a_{a0} - a_{ci} x_{i1} - \ldots \right.$$
$$\left. - a_{cj}(x_{ij} + \xi_{ij}) - \ldots - a_{cm} x_{im} \quad \right) + 2\gamma \xi_{ij} = 0 \tag{10}$$

Eq.(10) is arranged as:

$$\left(\gamma + \sum_{c=1}^{C} u_{ci}^{\theta} a_{cj} \right) \xi_{ij} = \sum_{c=1}^{C} u_{ci}^{\theta} a_{cj} \left(y_i - a_{a0} - a_{ci} x_{i1} - \ldots - a_{cm} x_{im} \right) = 0 \tag{11}$$

Therefore, the updating rule for ξ_{ij} is defined as:

$$\xi_{ij} = \frac{\sum_{c=1}^{C} u_{ci}^{\theta} a_{cj} \left(y_i - a_{c0} - a_{c1} x_{i1} - \ldots - a_{cm} x_{im} \right)}{\gamma + \sum_{c=1}^{C} u_{ci}^{\theta} a_{cj}^2} \tag{12}$$

Then, this tolerance value is calculated in each categorical variable.

By the way, the tolerance updating step should be withheld from the FCRM algorithm in the first several iterations because random initial regression models may make the tolerance values unstable. So, the tolerance values should be updated for adjusting stable prototypical regression models.

4 Numerical Experiment

The characteristic features of the proposed algorithm are demonstrated with an artificial data set shown in Fig.1. The objective variable y is dependent to two exploratory variable x_1 and x_2. As seen in Fig.1, two local regression planes $y=x_1+x_2$ and $y = -x_1 - x_2$ were generated. The artificial data set is composed of 60 objects (30 objects each in clusters), which are distributed along these two local regression planes with random perturbations. Then, x_2 is transformed into a categorical variable in such a way that 30 objects of each cluster are categorized into three categories with 10 objects each in ascending order as (Category 1: small, Category 2: medium, category 3: large). In this data set, the categorical observation x_2 has ambiguity caused by the coding into the three categories, i.e., loss of information. In this experiment, the goal of applying the proposed algorithm is to reconstruct the original local regression models of the fully numerical observations from the mixed data set by properly adjusting the quantified scores for categorical observations.

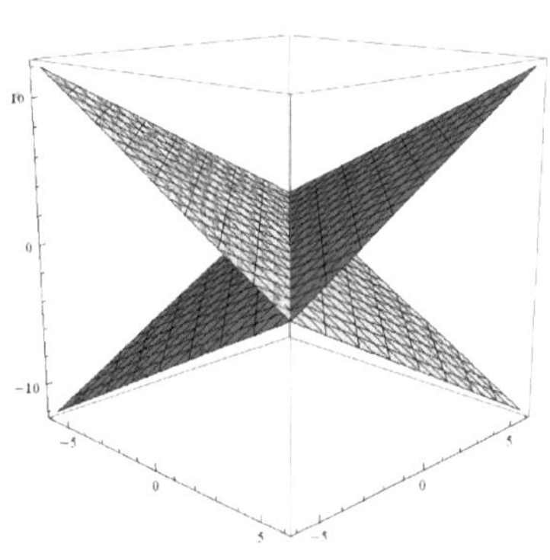

Fig. 1. Two Local regression planes ($y=x_1+x_2$ and $y=-x_1-x_2$)

The proposed new FCRM algorithm was applied to the data set and the result is compared with that of the conventional one without tolerance. Tables 1 and 2 show categorical quantified scores estimated by the optimal scaling mechanism. Tables 3 and 4 compare the prototypical regression models estimated in each cluster.

Table 1 implies that the conventional FCRM failed to properly reveal the intrinsic relation among categories by optimal scaling of them and the derived regression models shown in Table 3 are not the optimal ones. On the other hand, as shown in Table 2, the proposed method could properly reconstruct the mutual relation among three categories in ascending order. Then, the original linear generative models were successfully revealed as shown in Table 4. By the way, the regression coefficient on x2 (categorical variable) is much larger than that on x1 although they had similar values in the original generative models. It is because the optimal scaling mechanism normalized the quantified scores so as to have variance σ2, which is smaller than the original one, and properly re-adjusted their influences by a larger regression coefficient.

The result implies that the proposed method is useful for handling the ambiguity of categorical variables and performs robust fuzzy c-regression correctly.

Table 1. Quantified category scores based on optimal scaling in the conventional FCRM

Category	1	2	3
Score	0.141	1.138	-1.403

Table 2. Quantified category scores based on optimal scaling in the proposed method

Category	1	2	3
Score	-1.855	0.440	1.680

Table 3. Regression models based on optimal scaling in the conventional FCRM

Cluster	Regression model
1	$y=-4.093+0.629x_1-0.373x_2$
2	$y=4.805-1.939x_1-0.339x_2$

Table 4. Regression models based on optimal scaling in the proposed method

Cluster	Regression model
1	$y=0.643+0.726x_1+2.470x_2$
2	$y=-0.622-0.768x_1-2.650x_2$

5 Conclusion

In this paper, we proposed a robust fuzzy c-regression models based on optimal scaling considering ambiguity of categorical variables. By introducing tolerance for handling ambiguity of nominal categories, the FCRM algorithm was robustified. Experimental result demonstrated that the proposed method can reconstruct the properly quantified scores of categories, which suit the original generative model, even if the category information has ambiguity of nominal categories although the conventional FCRM algorithm failed to reveal the original generative models.

A potential future work is to apply the proposed method to a residential solar electric power analysis with the goal of revealing the intrinsic relation among solar electric power and meteorological observations through construction of multiple local regression models because some meteorological observations often given by categorical information with ambiguity.

Acknowledgements. This work was supported in part by the Ministry of Education, Culture, Sports, Science and Technology, Japan, under Grant-in-Aid for Scientific Research (23500283, 23500284).

References

1. Quandt, R.E.: The estimation of the parameters of a linear regression system obeying two separate regimes. Journal of the American Statistical Association 53, 873–880 (1958)
2. Quandt, R.E.: A new approach to estimating switching regressions. Journal of the American Statistical Association 67, 306–310 (1972)
3. Hathaway, R.J., Bezdek, J.C.: Switching regression models and fuzzy clustering. IEEE Trans. Fuzzy Systems 1(3), 195–204 (1993)
4. Bezdek, J.C.: Pattern Recognition with Fuzzy Objective Function Algorithms. Plenum Press (1981)
5. Young, F.W., De Leeuw, J., Takane, Y.: Regression with qualitative and quantitative variables: an alternating least squares method with optimal scaling features. Psychometrika 41, 505–529 (1976)
6. Takane, Y., Young, F.W., De Leeuw, J.: Nonmetric individual differences multidimensional scaling: an alternating least squares method with optimal scaling features. Psychometrika 42, 7–67 (1977)
7. Young, F.W., Takane, Y., De Leeuw, J.: Principal components of mixed measurement level multivariate data: an alternating least squares method with optimal scaling features. Psychometrika 43, 279–281 (1978)
8. Miyamoto, S., Ichihashi, H., Honda, K.: Algorithms for Fuzzy Clustering: Methods in c-Means Clustering with Applications. STUDFUZZ, vol. 229. Springer, Heidelberg (2008)
9. MacQueen, J.B.: Some methods of classification and analysis of multivariate observations. In: Proc. of 5th Berkeley Symposium on Math. Stat. and Prob., pp. 281–297 (1967)
10. Hamasuna, Y., Endo, Y., Miyamoto, S.: On tolerant fuzzy c-means clustering. Journal of Advanced Computational Intelligence and Intelligent Informatics 13(4), 421–428 (2009)

11. Endo, Y., Hasegawa, Y., Hamasuna, Y., Kanzawa, Y.: Fuzzy c-means clustering for uncertain data using quadratic penalty-vector regularization. Journal of Advanced Computational Intelligence and Intelligent Informatics 15(1), 76–82 (2011)
12. Honda, K., Uesugi, R., Ichihashi, H.: FCM-type fuzzy clustering of mixed databases considering nominal variable quantification. Journal of Advanced Computational Intelligence and Intelligent Informatics 11(2), 162–167 (2007)
13. Honda, K., Ohyama, T., Ichihashi, H., Notsu, A.: FCM-type switching regression with alternating least squares method. In: Proc. 2008 IEEE International Conference on Fuzzy Systems, pp. 122–127 (2008)

Combination of Genetic Algorithm and Ant Colony Optimization for QoS Multicast Routing

Bo Peng[1] and Lei Li[2]

[1] Graduate School of Engineering, Hosei University,
Koganei, Tokyo 184-8584 Japan
bo.peng.8k@stu.hosei.ac.jp
[2] Faculty of Science and Engineering, Hosei University,
Koganei, Tokyo 184-8584 Japan

Abstract. QoS-aware multicast routing service is becoming an important requirement of computer networks supporting group-based applications, such as multimedia conferencing, video conferencing, video telephony and distance learning. These real-time multimedia applications require the transmission of messages from a sender to multiple receivers subject to QoS constraints. This requires the underlying multicast routing protocol to find a QoS constrained minimum cost multicast spanning tree. However, the problem of finding the minimum cost multicast tree is known to be an NP -complete problem. In this paper, we present a new method GAACO to solve this minimum cost multicast routing problem. In this method, genetic algorithm (GA) and ant colony optimization (ACO) are combined to improve the computing performance. The simulation results show that the proposed GAACO algorithm has superior performance when compared to other existing algorithms.

Keywords: Ant Colony Optimization, Genetic Algorithm, QoS, Multicast Routing, Multicast Communication.

1 Introduction

With the growing demands of distributed multimedia applications, such as video on demand, multimedia conference, efficient and effective support of QoS has become more and more essential. Real-time multimedia applications require strict QoS guarantee on different parameters like delay, bandwidth, delay jitter, and packet loss ratio.

The routing of multicast traffic requires the construction of a multicast tree, on which a multicast source sends a single copy of data to a group of receivers instead of sending a separate copy of the data to each receiver as in a unicast communication. Thus, Multicast routing algorithms used to compute multicast trees that satisfy the QoS requirements of network applications have come under increased scrutiny in recent years. The main goal of multicast routing algorithm is to minimize the communication resources used by the multicast session. This is achieved by minimizing the cost of the multicast tree, which is the sum of the costs of the links in the multicast tree.

Y.I. Cho et al. (eds.), *Soft Computing in Artificial Intelligence*,
Advances in Intelligent Systems and Computing 270,
DOI: 10.1007/978-3-319-05515-2_6, © Springer International Publishing Switzerland 2014

GA is a part of evolutionary computing [1][2][3]. In early iterations, GA has fast convergence rate. When generation increases, GA may get stuck at a local optimum, and also the speed of convergence is slowed down. ACO was initially proposed to solve travel salesman problems (TSP) [4] [5]. Some ant colony algorithms have been proposed for solving multicast routing problem [6]. In this paper, we study the QoS multicast routing problem, and present a new multicast routing algorithm GAACO with multiple Qos constraints based on GA and ACO hybrid strategy. This algorithm took advantage of GA and ACO, and avoids the premature convergence of GA and low local optimization ability. Simulation results show that the algorithm is an effective approach to multicast routing problem with multiple QoS constrains.

2 Problem Description and Formulation

A multimedia communication network can be expressed as a weighted graph $G = (V, E)$, where V is the set of nodes representing routers, E is the set of links that connect the routers. Let $P_T(s, d_i)$ be the route from source node $s \in V$ to destination node d_i. $T(s, M)$ is a multicast tree. $M \in \{V - \{s\}\}$ is the set of destination nodes.

The total cost of multicast tree $T(s, M)$ is defined as the sum of the cost of all links in that tree and can be given by

$$cost(T(s, M)) = \sum_{e \in T(s,M)} Cost(e) \tag{1}$$

$P_T(s, d_i)$ is the routing path between source node s and destination node d_i of multicast tree $T(s, M)$, which has following relations:

$$Delay(P(s, d)) = \sum_{e \in P_T(s,d_i)} Delay(e) \tag{2}$$

$$Bandwidth(P(s, d)) = min_{e \in P_T(s,d_i)} \{Bandwidth(e)\} \tag{3}$$

$$Packet\_loss(P(s, d)) = 1 - \prod_{e \in P_T(s,d_i)} [1 - Packet\_loss(e)] \tag{4}$$

$$Dealy\_jitter(P(s, d)) = \sum_{e \in P_T(s,d_i)} Delay\_jitter(e) \tag{5}$$

We assume that the delay constraint is D_{max}, bandwidth constraint is B_{min}, delay jitter constraint is DJ_{max}, and packet loss ratio constraint is PL_{max}. The QoS multicast routing problem is defined as minimization of the cost function $min\{cost(T(s, M))\}$, subject to the following conditions:

1. $Delay(P_T(s, d_i)) \leq D_{max}$
2. $Bandwidth(P_T(s, d_i)) \geq B_{min}$
3. $Dealy_{jitter(P_T(s,d_i))} \leq DJ_{max}$
4. $Packet\_loss(P_T(s, d_i)) \leq PL_{max}$

3 Proposed GAACO Algorithm for QoS Multicast Routing

In GAACO, the number of gene in each chromosome equates to the number of the destination nodes, and also equates to the number of ants in sub-colony, and each chromosome and sub-colony can generate a multicast tree.

The basic process of GAACO is: First, during the network initialization phase we concern about the bandwidth constraint. We delete all the links, which their bandwidth are less than the minimum required bandwidth. Then, we apply GA to find the least cost multicast tree. GA procedures are iterated until a pre-specified termination criterion is satisfied, and then we combine ACO. To enhance the search capability at the beginning stage of ACO, we update the initial pheromone according to the fitness value of the population generated by GA. ACO procedures are iterated until stopping criteria is satisfied.

3.1 Encoding and Population Initialization Method

In this algorithm, every chromosome represents a multicast tree that covers source node and destination nodes. The creation of initial population is based on Dijkstra kth shortest path algorithm. For each destination node $d_i \in M$, we compute least-cost paths from s to d_i by using Dijkstra kth shortest path algorithm to construct a candidate paths set. Let Q_i be the candidate paths set for destination node d_i:

$$Q_i = \{P_i{}^1, \dots, P_i{}^j, \dots, P_i{}^k\} \tag{6}$$

A multicast tree $T(s, M)$ is encoded as an array of $m = |M|$ elements, where each element is a path from source node s to a destination node $d_i \in M$, i.e., $T(s, M) = \{p_1, p_2, \dots, p_m\}$, where $p_i = p(s, d_i)$, $d_i \in M$, p_i is a path selected from the candidate paths set of destination node d_i. The relation between chromosome, gene, and candidate paths set is shown in Fig.1.

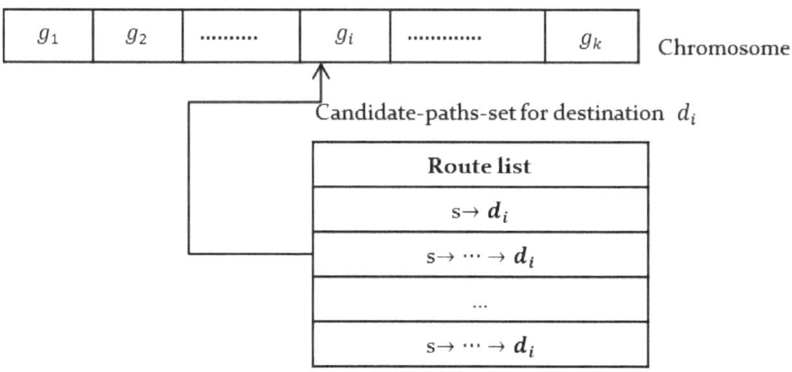

Fig. 1. Chromosome structure

3.2 Fitness Function

In this paper, we define the fitness function for each individual, the multicast tree $T(s,M)$, using the penalty function, as follows:

$$f(T(s,M)) = f_c \times f_d \times f_{dj} \times f_{pl} \tag{7}$$

$$f_c = \frac{\alpha}{cost(T(s,M))} = \frac{\alpha}{\sum_{e\in T(s,M)} Cost(e)} \tag{8}$$

$$f_d = \prod_{d_i \in M} \delta_d\big(Delay(p(s,d_i)) - D_{max}\big) \tag{9}$$

$$f_{dj} = \prod_{d_i \in M} \delta_{dj}\left(Delay_{jitter(p(s,d_i))} - DJ_{max}\right) \tag{10}$$

$$f_{pl} = \prod_{d_i \in M} \delta_{pl}(Packet\_loss(p(s,d_i)) - PL_{max}) \tag{11}$$

$$\delta_d = \begin{cases} 1, & T \le 0 \\ r_d, & T > 0 \end{cases} \tag{12}$$

$$\delta_{dj} = \begin{cases} 1, & T \le 0 \\ r_{dj}, & T > 0 \end{cases} \tag{13}$$

$$\delta_{pl} = \begin{cases} 1, & T \le 0 \\ r_{pl}, & T > 0 \end{cases} \tag{14}$$

where α is a positive real coefficient, $\delta(T)$ is the penalty function. When the individual satisfies the QoS constraint, its value is 1, or else r $(0 < r < 1)$. The value of r determines the degree of penalty. In this algorithm, we select $r_d = r_{dj} = r_{pl} = 0.5$ in our simulation experiments.

3.3 Genetic Operators

We use the roulette wheel selection method, one point crossover and mutation operators as genetic operators.

The mutation operation is performed according to the probability of mutation p_m. First, a mutation gene g_i is selected randomly. Then, the gene g_i is replaced with a new candidate path p_i, which is selected randomly from the candidate paths set of the destination node d_i.

3.4 Multiple ant Sub-colonies

We divided the ant population into equally-sized sub-colonies, and in each sub-colony the number of ants is equal to the number of destination nodes. Each sub-colony searches for a multicast tree.

In some proposed ACO algorithms for multicast routing, every ant has its own definite destination node while searching for a path. However, in large-scale networks this method may take a long time to find the definite destination node. Therefore, in this paper every ant starts to explore all available paths without a given destination node.

3.5 ACO Algorithm State Transition Rule

Each ant chooses the next node to move, considering the trail intensity and cost between the two nodes. An ant positioned in node i chooses the node j for its next step according to the following rule:

$$j = \begin{cases} argmax_{j \in allowed_k(t)} \left[\tau_{ij}(t)\right]^{\alpha} \left[\eta_{ij}\right]^{\beta}, q \leq q_0 \\ J \qquad , q > q_0 \end{cases} \tag{15}$$

where $\tau_{ij}(t)$ is the amount of pheromone trail between node i and j at time t, η_{ij} is the heuristic function which is the inverse of cost between node i and j, α and β are two parameters which determine the relative importance of the pheromone trail and the heuristic information, q is a random number uniformly distributed in $[0,1]$, q_0 is a pre-specified parameter, $allowed_k(t)$ is the set of feasible nodes currently not yet visited by ant k at time t. With probability $1 - q_0$ the next node j is chosen according to the probability distribution given by Eq.16.

$$P_{ij}^{k}(t) = \begin{cases} \frac{\left[\tau_{i,j}(t)\right]^{\alpha} \left[\eta_{i,j}\right]^{\beta}}{\sum_{s \in allowed_k(t)} \left[\tau_{i,s}(t)\right]^{\alpha} \left[\eta_{i,s}\right]^{\beta}}, j \in allowed_k \\ 0 \qquad , otherwise \end{cases} \tag{16}$$

3.6 Global Updating Strategy

Global pheromone updating rule is performed after all ants have constructed their paths. In this paper, we adopt the rank-based version of Ant System $(ASrank)$ [7]. In each iteration the solutions generated by all sub-colonies are ranked from 1 to m according to the tree costs of their solutions $(C_1(t),\ C_2(t), \ldots, C_m(t))$. Only the best $(w-1)$ solutions of each iteration are allowed to deposit pheromone. The r-th best sub-colony of the current iteration updates with weight $(w-r)$. The global best solution, which gives the strongest feedback, updates with weighed coefficient w. The global updating rule is

$$\tau_{ij}(t+1) = (1-\rho) \cdot \tau_{ij}(t) + \sum_{r=1}^{w-1} (w-r) \cdot \Delta\tau_{ij}^{r}(t) + w \cdot \Delta\tau_{ij}^{gb}(t) \tag{17}$$

$$\Delta\tau_{ij}^{r} = \begin{cases} \frac{Q_1}{cost_r}, if\ link\ (i,j)\ is\ visited\ by\ sub-colony\ with\ rank\ r \\ 0 \quad , otherwise \end{cases} \tag{18}$$

$$\Delta\tau_{ij}^{gb} = \begin{cases} \frac{Q_1}{cost_{gb}}, & if\ link(i,j) \in global\ best\ solution \\ 0, & otherwise \end{cases} \tag{19}$$

where Q_1 and w are constants, $cost_{gb}$ is the cost of global best solution in the searching history, and $cost_r$ refers to the cost of tree constructed by k-th sub-colony with rank r in current iteration.

3.7 Limited Pheromone Trails

To avoid search stagnation and premature convergence, we impose upper and lower bounds on pheromone trails based on Max-Min Ant System (MMAS) [8], so that all pheromone trails strengths are limited to the interval $[\tau_{min,}\ \tau_{max}]$.

$$\tau_{max} = \frac{Q_2}{\rho \cdot cost_{gb}} + \frac{Q_2 \cdot (w-1)}{cost_{gb}} \tag{20}$$

$$\tau_{min} = \frac{\tau_{max}}{m} \tag{21}$$

4 Simulation Results and Analysis

In this section, we have used the simulation results to compare the performance of the proposed GAACO algorithm with Younes-ACO[6], GA-Wang[9] and AGSAA[11].

The experimental simulations are achieved by MATLAB 7.11.0 to implement both algorithms. All simulation experiments are run on a personal computer (Intel Core i7 Processor (3.4GHz), 8.00GB RAM). The network topology is created by Salama graph generator in this experiment [10]. The multicast group is randomly selected in the graph. The numbers of network nodes are 20, 40, 60, 80 and 100 nodes respectively, and the size of multicast group is 5.

In simulations the following parameter settings are achieved: $NC_{max} =$ 500(iteration numbers), $m = 10$, $\alpha = \beta = 1$, $\rho = 0.1$, $Q_1 = 4$, $Q_2 = 10$, $w = 6$. For GA, we set the size of population $N_{pop} = 10$, crossover probability $p_c = 0.9$, mutation probability $p_m = 0.1$. Dijkstra *kth* shortest path algorithm was designed to automatically generate the shortest 20 paths for each destination node.

Fig. 2 shows the average tree cost with the change of number of generations in 60-node network. As shown in this figure, our proposed algorithm GAACO shows significantly improved computational result than other algorithms. It can be seen that GAACO has high speed of convergence and better global searching capability than other algorithms.

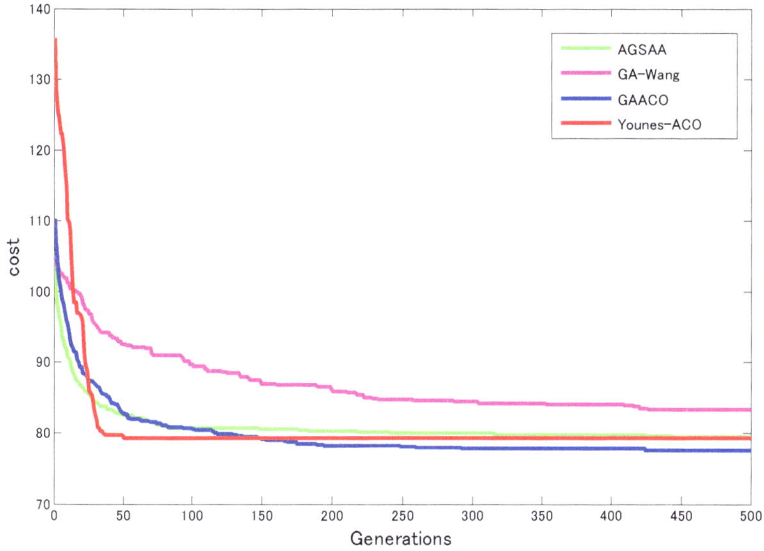

Fig. 2. Comparison of average cost performance in 60-node network

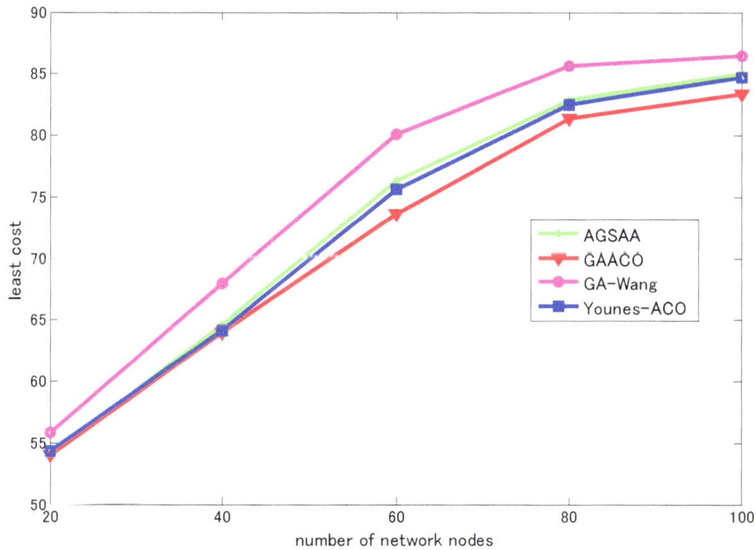

Fig. 3. Comparison of least tree cost in different network scale

Fig. 3 shows the comparison of least tree cost in different network scale. It can be seen that in small networks, the average least costs of each algorithm are almost same, but in large-scale networks GAACO has good performance of cost than other algorithms.

5 Conclusion

In this paper, we presented a method for least-cost QoS multicast routing based on GAACO that combined GA and ACO. In this algorithm, the number of gene in each chromosome equates to the number of the destination nodes, and also equates to the number of ants in sub-colony, and each chromosome and sub-colony can generate a multicast tree.

The simulation results show that this hybrid algorithm has not only fast convergence speed and good performance of cost, but also is an effective solution to the least-cost QoS multicast routing problem in large-scale networks.

Acknowledgements. The author would like to express gratitude to all those who have helped her during the writing of this paper. She deeply appreciates the contribution to this paper made in various ways by her friends and classmates.

References

1. Holland, J.H.: Adaptation in Natural and Artificial Systems. The University of Michigan Press, Ann Arbor (1975)
2. Goldberg, D.E.: Genetic Algorithms in Search, Optimization and Machine Learning, p. 432. Addison Wesley (1989)
3. Forrest, S., Mitchell, M.: Relative building-block fitness and the building-block hypothesis. In: Whitley, L.D. (ed.) Foundations of Genetic Algorithms 2. Morgan Kauffman, San Mateo (1993)
4. Colorni, A., Dorigo, M., Maniezzo, V.: Distributed optimization by ant colonies. In: Proceedings of ECAL 1991-European Conference on Artificial Life, pp. 134–142 (1991)
5. Colorni, A., Dorigo, M., Maniezzo, V.: An investigation of some properties of an ant algorithm. In: Proceedings of the Parallel Problem Solving from Nature Conference, pp. 509–520 (1992)
6. Younes, A.: An Ant Algorithm for Solving QoS Multicast Routing Problem. International Journal of Computer Science and Security (IJCSS) 5(1), 156–167 (2011)
7. Bullnheimer, B., Hartl, R.F., Strauss, C.: A new rank-based version of the ant system: a computational study. Central European Journal of Operations Research 7(1), 25–38 (1999)
8. Stützle, T., Hoos, H.: MAX-MIN Ant System and Local Search for the Traveling Salesman Problem. In: Proceedings of the IEEE International Conference on Evolutionary Computation (ICEC 1997), pp. 309–314 (1997)
9. Wang, X.H., Wang, G.X.: A multicast routing approach with delay-constrained minimum-cost based on genetic algorithm. Journal of China Institute of Communications 23(3), 112–117 (2002)
10. Salama, H.F., Reeves, D.S., Viniotis, Y.: Evaluation of multicast routing algorithms for real-time communication on high-speed networks. IEEE Journal on Selected Areas in Communications (1997) 15(3), 332–345 (1997)
11. Peng, B., Li, L.: A Method for QoS Multicast Routing Based on Genetic Simulated Annealing Algorithm. International Journal of Future Generation Communication and Networking 5(1), 43–60 (2012)

Building a Knowledge Base for Expert System in Information Security

L. Atymtayeva, K. Kozhakhmet, and G. Bortsova

Department of Computer Engineering,
Kazakh-British Technical University,
Tole bi st., 59,
Almaty, Kazakhstan
{l.atymtayeva,kanik85,gerdabortsova}@gmail.com

Abstract. In the days of technological advancement, a role of information security (IS) is very important. There is an urgent need in implementing and assessing information security at a good level. However, it is accompanied with very high costs: experts in IS are quite expensive specialists. An automation of some security implementation and evaluation tasks can reduce these costs and potentially increase the quality of IS strategies being developed and IS audit quality. We believe that expert systems approach can be beneficial in achieving this automation. Though information security is a very broad field, encompassing many complex concepts, we are trying to develop a methodology of formalizing of IS knowledge to build a knowledge base for expert system that can serve as IS audit expert. In this paper we discuss methods for knowledge base building.

Keywords: expert systems, information security, knowledge base.

1 Introduction

Nowadays it is almost impossible to find a branch of human activity where there is no information technology (IT). Because of IT technologies rapid growth the companies often encounter with the need of increasing the information security. However, information security is a comprehensive system that is very difficult to manage. As a consequence, in the most organizations there is a risk of the information system safety.

The best solution in these circumstances may become the audit of the information security in organizations. The audit process is highly expensive in terms of time and cost as well as in the degree of involvement of human resources.

One of the efforts taken in reducing expenses and facilitating audit is the use of special tools such as checklists and questionnaires, to identify gaps between certain security standards and existing organization's security practices. ISO 17799 Checklist ([1]) provides number of audit questions regarding ISO standard guidelines. ISO IEC 27002 2005 (17799) Information Security Audit Tool, described in [2], offers several hundred audit questions (questions are stated in yes-no form), pointing to security practices that need to be implemented and actions that should be taken (in case of

Y.I. Cho et al. (eds.), *Soft Computing in Artificial Intelligence*,
Advances in Intelligent Systems and Computing 270,
DOI: 10.1007/978-3-319-05515-2_7,

“no” answer to question). Though these tools cannot be used independently, without any additional security measurements, they still are useful for human auditors. Thus, auditing process can be seen as a process of asking questions and making conclusions from answers.

Another effective tool for the audit is to develop a knowledge base that will provide information for Chief Information Security Officers (CISOs) and will help them to find the right management decisions on the information security policy [3]. Key components of the knowledge base are: "Asset", "Source" (standard), "Vulnerability," "Step" (a refinement of the part of “Guideline” in a special section of the standard) and others.

Every "Step" refers to the protected Object, to the type of Vulnerability it is against as well as to the cross-references to other stored Guidelines. This tool provides search-based knowledge management directives, standards, analysis of the components and issuing recommendations. As a result, the so-called meta-model of the security standard recommendations could be constructed[4-7].

By the reason of highly expensive process of information security Auditing in terms of high cost of different resources (time, people, expenses) the reducing the cost of the audit process is a priority for any organization. Automating the audit process by creating intelligent software (expert system) can significantly reduce costs, since the main work on decision-making is carried out automatically, based on computer analysis of the situation and issuing guidelines and recommendations.

We think that expert systems have much to offer in the case of IS audit automation. Expert systems (ES) approach firstly fits question-answer format of auditing, secondly, ES function on the basis of meta-model that reflects knowledge in target field. Emulating the way expert in particular field thinks and implementing common human logic can give a system that is able to assess the situation and make decisions.

Previously expert systems approach in security area was applied in computer security auditing. An Expert System in Security Audit (AudES), designed for automating some audit procedures, like identifying potential security violations by scrutinizing system logs, is described in [8].

But the application of the methodology of expert systems in IS auditing in the broadest sense (not only in computer security) (what actually we would like to realize) remains largely untouched. Our task is to study and solve the problems of development of expert systems in a wide range of information security audit, which includes aspects of computer security.

In the process of the development of the ES in IS, we passed the following stages:

1. Building a high-level structure of the knowledge base for IS. This stage encompasses analyzing IS standards and deriving key concepts from them; formalizing a process of IS assessment and decision-making.

2. Development of system workflow. At this stage, we developed a model where all parts of our system (key elements, identified on the previous stage) can nicely play together [4-5].

3. Developing a methodology for population of the knowledge base. The stage involves deriving lower-level concepts and subconcepts.

In this paper, we focus on the last stage, but for good understanding it's necessary to stop at the previous two.

2 Knowledge Base Structure and System's Workflow

The aim of the expert system's knowledge base is to represent domain specific knowledge in the form which can be used by a computer to effectively operate on this knowledge. In order to meet this requirement, we have chosen to build ontology (knowledge model) of information security domain. Our ontology consists of 4 main entities and relationships between them.

The ontology is divided into two parts: the concepts representing IS domain knowledge (which actually are core concepts of the domain) and the concepts representing concrete information about considered organization, which are essential in measurement of its security level. These concepts are:

-*Threat* is a potential cause of an unwanted incident, which may result in harm to a system or organization [ISO].

-*Vulnerability* is a physical, technical or administrative weakness which could be exploited by threats.

-*Control* concept is used to mitigate vulnerabilities by implementing either organizational or physical measures.

-*Asset* is anything that has value to the organization [ISO]. Also assets are used to implement controls.

The most important relations between these concepts are:

-Threat threatens asset.
-Vulnerability is exploited by threat Severity.
-Vulnerability is mitigated by control.
-Control is implemented by asset Effectiveness.
-Asset has vulnerability.

This is a short explanation of main components of our knowledge model, on top of which we build a knowledge base. We also develop a methodology of risk assessment using the concepts and relations in the knowledge base [1].

The next our step was to determine how our system uses described knowledge model to assess information security level in given organization. In a nutshell, our system asks the user (a person responsible for security, organization's management or company's employees) a set of questions of various forms, analyzes the answers to questions using its knowledge base, and calculates security risks together with giving recommendations for improving security level.

The workflow of our system encompasses the following steps:

1. Collecting Enterprise Data

The first stage in the work of our system is collecting data of an organization to be audited. It includes gathering knowledge about assets (everything that need to be protected) and their value to the company.

To achieve this, a variety of question templates is constructed. Basically, the following information is expected as an answer to these questions:

-asset (assets) that is (are) presented / not presented;
-and/or value of assets (their attributes);
-the dependencies between the different assets.

Example of possible questions:

Do you have any *data centers* (asset)? Where they are located? (The possible answer may be a plan of company's buildings, which allows revealing dependencies of the asset on other assets, e.g. a dependency of a data center on the building it is situated in, the building's heating system, fire protection system, etc.) What would be a loss in case your data center was damaged or destroyed? (The answer should be in money or in per cent equivalent which represent the value of asset)

We are certain about creating as full and detailed categorization of assets and types of dependencies between them as possible, as it is a key to our results' accuracy. This categorization lays the ground to creation of the set of questions for this stage.

2. Threats, Vulnerabilities and Controls Identification

After the system has collected all necessary information about the target organization, it tries to find threats relevant to assets (threats which threaten assets) identified at the previous stage. Then it tries to find vulnerabilities that can exploit these threats and controls that could possibly mitigate these vulnerabilities.

The reason we collect data about these three concepts (asset, threat vulnerability), besides construction of a complete information security domain ontology, is risk calculation. We will explain it in more detail further.

3. Collecting Data about Organization's Security Policy and Its Implementation

In order to assess the quality of implementation of a control in the target organization, the system asks several questions regarding each of the standards' control. Each question should identify a value of one variable. For example:

- How often do you perform backups of sensitive data? (The answer to the question is a value of a variable called backup frequency expressed in backups/year)
- How many percent of sensitive information do you usually backup? (The variable is backup coverage in %)

As a concluding phase of the whole process, on the basis of these values and using fuzzy/non-fuzzy rules, a control effectiveness is calculated (i.e. how effective is particular control's implementation in the organization)[9-12].

To be clearer we can summarize above information into the scheme on the figure below (Figure 1).

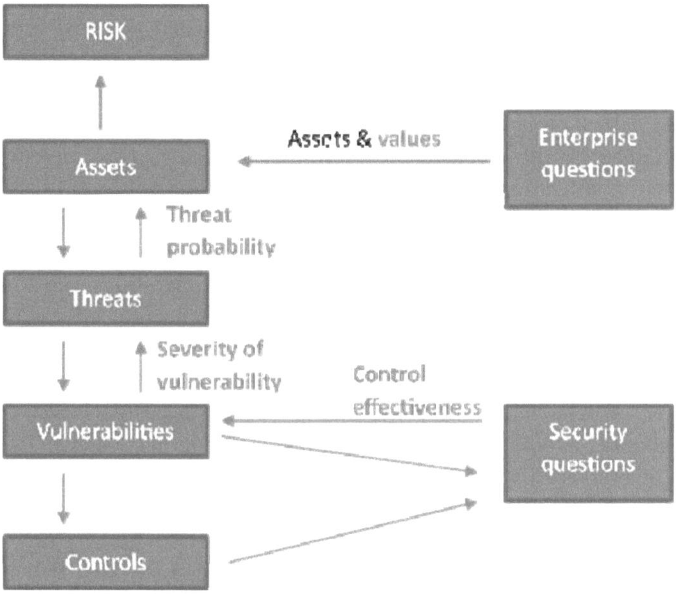

Fig. 1. Scheme of mapping process for all components of ontology

"Enterprise questions" is a set of questions about target organization, the answers to which (list of assets, dependencies, and values of assets) are captured in the very first stage. It results in a populating enterprise part of the knowledge base with asset instances. The system then finds appropriate threat, vulnerability, control instances and forms new set of questions regarding security practices ("Security questions") in the given company. The answers to these questions allow firstly to calculate control effectiveness of each control, secondly use severity of vulnerability coefficients (pre-defined) when the threat probability coefficients for each threat can be derived, and, finally, calculate security risks on the basis of latter coefficients and asset values.

However, to make the system working, we need a rich set of objects (concepts' instances) to operate on. We propose using international information security standards (like ISO 27002) for extraction of the knowledge about the field, particularly expressed in a form of elements of our ontology, i.e. instances of 4 concepts and their relationships.

3 The Method to Collect Concepts' Instances

The subjects of information relations (source, the owner, the owner or the user information) define a set of data resources to be protected against various attacks. For assets information systems usually include: material resources, information resources such as analytical, service, control information at all stages of the life cycle: creation, processing, storage, transfer, disposal;

Information technology life cycle processes automated systems, providing information services, and etc [13].

The attacks are the result of the threats made through the various security vulnerabilities, and have a probability (in terms of risk of attack). Main security breaches the following concepts: the disclosure of information values (loss of confidentiality), unauthorized modification to their loss of integrity, or unauthorized loss of access to these values (loss of availability). In the result of analysis of security vulnerabilities, properties, sources of threats such as nature of the occurrence, character and attitude to the objects of Information Systems, and possible probabilities implementation in a particular environment, we can determine risks for given set of information resources [14-18]. This determination allows us to define the security policy. Elaborated subject of information relations strategy of protection may provide for each of the threat the possible line of behavior, namely, the attempt to eliminate the source of the threat; threat avoidance; adoption of threats; minimization of the damage from an attack caused by this threat by using the services and security mechanisms. It should be taken into account that individual vulnerability can be preserved after use of protective activities [19-21].

Proceeding from the given principles above, we can say that the modeling and classification of the sources of threats and their displays may be carried out by analysis of interaction of logical chains. These logical chains may be constructed by using security policy and analyzing the possible risks.

At the same time the security policy defines a coherent set of mechanisms and security services, adequate protection of the values and the environment in which they are used.

Thus, the process of providing information security should have comprehensive basic approach that is based on a deep analysis of possible negative and positive impact effects [22]. Such an analysis involves mandatory identification of possible sources of threats, the factors contributing to their display (vulnerabilities) and, as a consequence, the determination of the actual threats for information security. We can represent this logical chain as following:

Source of threat -> Threat -> Vulnerability -> Implementation of Threat - > Effect (Damage).

Relying on concepts introduced above, we can construct the following detailed ontological scheme for the issues of information security which contains the general relations between Threat, Vulnerability and Assets (named as information sources) and their subjective derivatives (figure 2).

3.1 The Classification of Information Security Threats

Providing of security of information is impossible without systematic analysis of the relevant security threats. Basis of such analysis should be classification of threats to certain basic features that gives to researcher (Expert in Information Security) general holistic view of the various variants of destructive influences and their impacts.

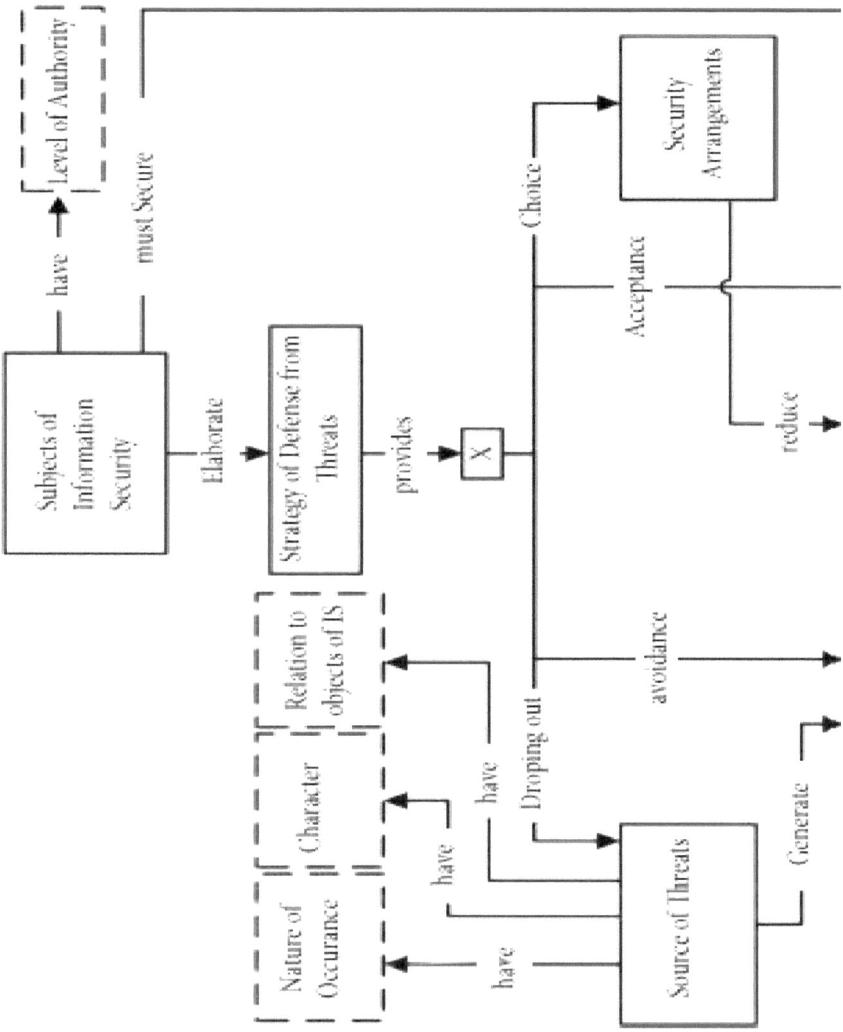

Fig. 2. Detailed ontological scheme for the issues of information security

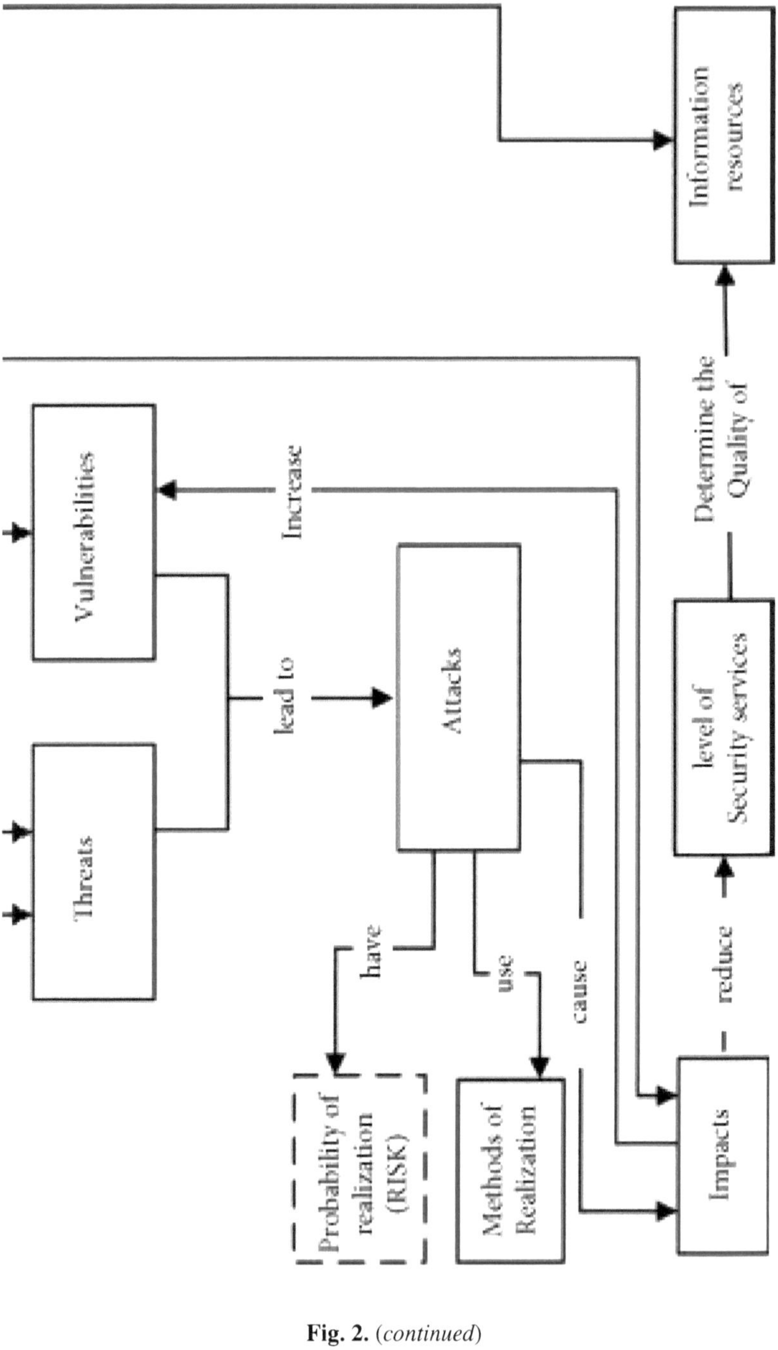

Fig. 2. (*continued*)

The literature review proposed number classifications of security threats showing various aspects of this problem [23 - 27]. However, being designed for a narrow range of specific tasks, they can't be the basis of the total ordering threats and highlight their most significant attributes for later synthesis and decomposition that can be effectively used in developing Knowledge Base Ontology of Expert System.

It is necessary to develop a generalized classification, which must allow considering several characteristics of threats as a subject of scientific research and later describe and show all possible types and derivatives. At the same time the classification of the factors influencing on information security should be conducted with the following requirements:

-sufficiency of the levels of classification factors allowing them to form a complete set;

- enough flexibility of classification, allowing expand set of classified factors and signs groups and make the necessary changes without disrupting the structure of the classification.

Under the security threat, as noted earlier, we can understand the situation, in which may be violated common services, such as integrity, confidentiality and availability of information. Morphological analysis shows that we can highlight the following basic components of information security threats: the source impact on the information system, the method exposure, the impact of information objects, as well as the results (or damages).

These statements while developing for the classification can be selected as the basic classification criteria for onward decomposition.

According to the standard [28], the factors affecting on information security can be classified by using their relationship of nature as objective and subjective, and by relation to the objects of information systems as internal and external. The general scheme of information security threats classification is shown in Figure 3.

The dividing of sources on subjective and objective is warranted on the basis of considerations to determine the guilt for the damage information. Dividing into internal and external sources is warranted because it can be different for the same methods of parrying threats to internal and external sources.

Moreover, both external and internal sources can be both intentional and unintentional.

Unintentional threats arise regardless of the will and the desire people. This type of threat is most often associated with direct natural or anthropogenic impacts on the physical elements of the information system and leads to malfunction of the system and / or physical damage (destruction) carriers, data processing and data transmission telecommunication channels.

Intentional threat in contrast to unintentional can be created only by people acting targeted to disrupt the work of an information system. Intentional threats may be divided into passive and active.

Passive threats are related to unauthorized access to information, without any amendment. Active threats associated with attempts to change (interception, modification, destruction) of the information or attempts to disable access to the information resources of legitimate users.

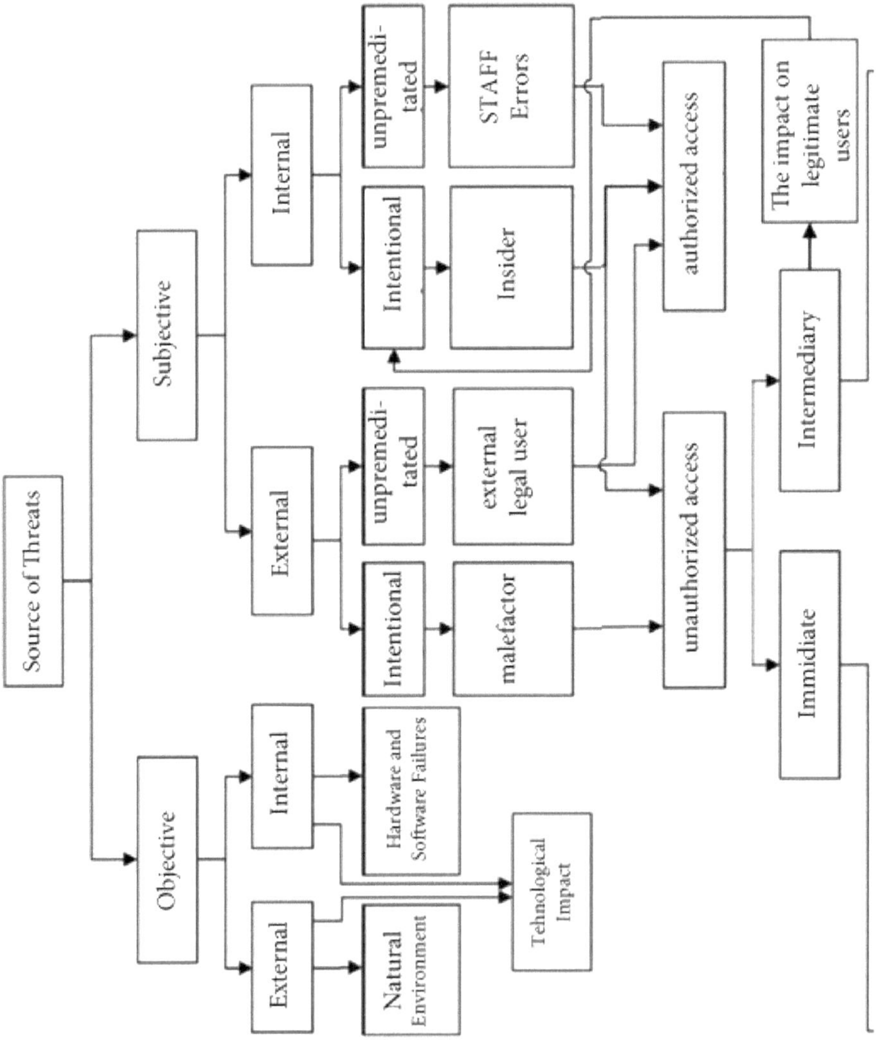

Fig. 3. The general scheme of information security threats classification

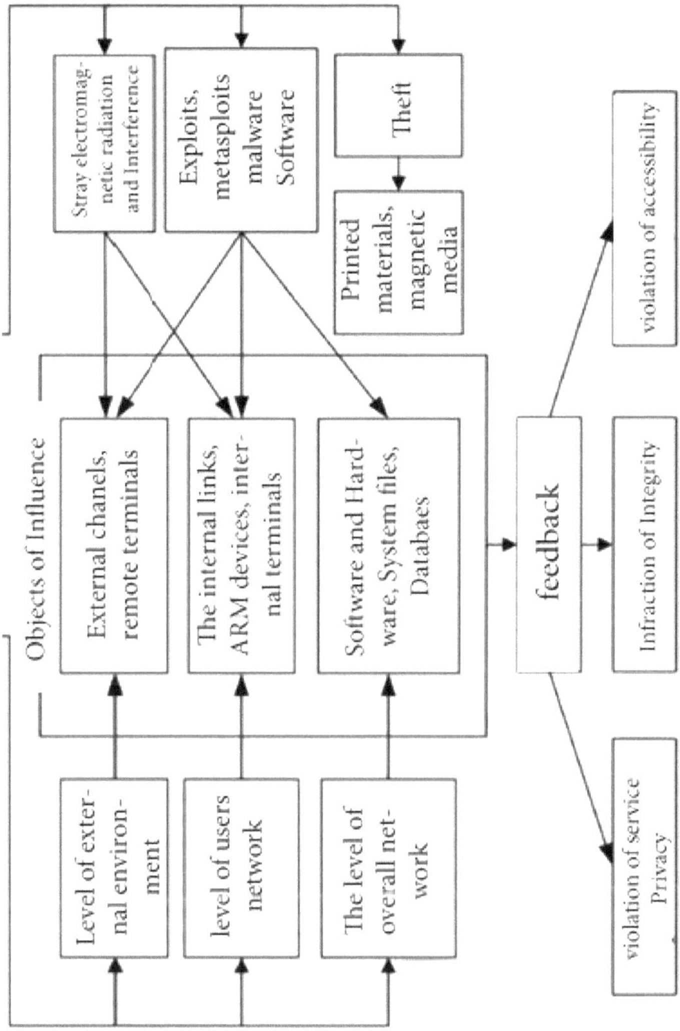

Fig. 3. (*continued*)

We can distinguish between several sources of threats:

Anthropogenic sources of threats. Anthropogenic sources of information security threats is the subject whose actions can be qualified as intentional or accidental violations. This group is the most extensive and is of most interest from the point of view of the organization of protection, as actions of the subject can be evaluated to predict and take adequate protection measures. Countermeasures in this case are directly controllable and depend on the will of the organizers of information security. As a source of anthropogenic threats we can consider subjects that have an access (authorized or unauthorized) to the work with standard means securable. Subjects (sources), whose actions may lead to a breach of information security can be both external and internal. External actors, sources of threat, in turn, can be accidental or intentional, and may have different levels of qualifications.

Below we enumerate some examples of external sources:

- criminal organizations;
- potential criminals and hackers;
- unscrupulous partners;
- technical staff telematics service providers;
- representatives of the supervisory organizations and emergency services;
- representatives of power structures.

Internal actors (sources), usually are highly qualified experts in the field of development and operation of software and hardware, are familiar with specificity of tasks, structure, basic features and principles of work software and hardware information security, and are able to use of standard equipment. For example:

- key personnel (users, programmers, designers);
- representatives of the service protection;
- support staff (cleaners, security);
- technical staff (maintenance engineering networks).

Technogenic sources of threats. The second group contains the sources of threats, determining the implications technocratic human activities that can get out of it control and stand on their own. These sources of threats are less predictable directly depend on the properties of art and therefore require the special attention.

Natural sources of threats. The third group of sources of threats combines circumstances components of an irresistible force, that there are circumstances that have an objective and absolute nature, which applies to all. Overwhelming power in the legislation and contractual practices include natural disasters or other circumstances that can't be provide [29]. Natural sources of potential threats to information security are typically external to the protected object and they relate mostly to the natural disasters: fires, earth-quakes, floods, hurricanes, and various contingencies circumstances, unexplained phenomena and other force majeure circumstances.

3.2 Classification of Vulnerabilities

Vulnerability inherent in the object information, inseparable from it and are caused by deficiencies in the process of operation, properties of architecture automated systems, communication protocols, and interface used by software and hardware platform, operating conditions and location, and etc.

Vulnerabilities may be presented in both the software and hardware, and organizational and legal support to information security. The main part of the vulnerabilities of the organizational and legal support contains the lack of regulations on businesses, and issues relating to information security [30-31]. An example of the vulnerability of this type is the lack of organization in the approved concept or information security policy, which would define the requirements for Protection of Information Systems, as well as specific ways to implement them.

Vulnerabilities in software and hardware can be presented in the software or hardware components of user workstations, information systems servers and communication equipment and communication channels for information security. Sources of threats can exploit vulnerabilities to break safety information to obtain illegal benefits (damage proprietor, owner, user information). Furthermore, there may be not malicious relation to sources of threats to intensify certain vulnerabilities that could be harmful.

Each threat can be compared by various vulnerabilities. Eliminating or substantially reducing the vulnerability affects is the ability to implementation of information security threats.

There are different approaches to systematizing vulnerabilities of information systems and technologies. In [31], for the convenience of the vulnerability analysis they are divided into classes, groups and subgroups. Thus the information security vulnerabilities may be divided as the following:

- objective vulnerabilities;
- subjective vulnerabilities;
- occasional vulnerabilities.

Objective vulnerabilities. Objective vulnerabilities depend on features of construction and technical characteristics of equipment used on the protected object. Full removal of these vulnerabilities is not possible, but they can significantly weaken the technical and engineering methods countering threats to the security of information. We can consider[32]:

- Related technical means of radiation (electromagnetic, electrical, sound);
- clickable vulnerabilities (hardware and software tab);
- vulnerabilities determined by the characteristics of the elements (elements that have electro-acoustic transducers or exposed electromagnetic fields);
- vulnerabilities determined by the characteristics of the protected object (location object, organization communication channels).

Subjective vulnerabilities. Subjective vulnerabilities depend on the actions of employees and, in general, may be eliminated by organizational software and hardware methods. These vulnerabilities include[33]:

- Errors (during the preparation and use of the software, with management of complex systems, with the use of equipment);
- Violations (regime of protection and protection, the mode of operation of technical means modes of use and privacy mode, etc.).

Occasional vulnerabilitie. Occasional vulnerabilities depend on the characteristics of environmental protected object environment and unforeseen circumstances. These factors are usually a little predictable, and their removal is only possible during the range of organizational and engineering activities for decreasing or mitigating threats to information security. We can consider the following types of vulnerabilities:

- Faults and failures (failures and malfunctions of technical equipment, aging and demagnetize media, software failures, failure of power supply, etc.);
- damage (lifeline enclosing structures, etc.).
  and etc.

4 Situational Sample

Let us look at sample of security state of any university [33-36]. In the modern university a huge number of various data related not only to the educational process, but to research and engineering studies, personal information of students and staff, service, commercial and other confidential information are stored and processed constantly. The growth of crime in the area of high technologies dictates its own requirements of resource protection of computer networks of educational institutions and sets the task constructing its own integrated security system. Its solution presupposes the existence of the legal framework, the formation of the concept of security, development activities, plans and procedures for the safe work, design, implementation and maintenance of technical equipment protection of information (EPI) within the educational institution. These components define single security policy information at the university. Specificity of information security in the educational system is that the university - a public institution with a volatile audience, but also a place increased activity of "beginners cyber criminals." The main group of potential offenders is students and some of them may have a high level of training. Age range is from 18 to 23 years that contribute to youthful maximalism. The potential threats to information security may be caused by the circumstances that some young people show off in front of fellow students knowledge in arrange virus epidemic, and gain administrative access to "punish" the teacher, block access to the Internet, etc [34]. Suffice it to recall that the first computer offenses were born at the university (example is the Morris worm).

Threats to information security and risk analysis of the university. For information risk analysis we should carry out the following activities:

- classify the objects to be protected, and to rank them in order of importance;
- determine the attractiveness of protection for attackers;
- identify possible threats to information security;
- consider possible ways to implement them (vulnerabilities);
- assess the damage from potential attacks on information resources.

We can distinguish the main objects of the university that are needed to protect:

- Accounting LAN data Planning and Finance Department, as well as
- statistical and historical data;
- database servers;
- Management Console user accounts;
- www / ftp server;
- LAN servers and research projects.

According to the above classification, taking into account the considered features we can study the following threats to information assets of the university (We give only some examples of threats):

Unintentional subjective Information Security threats:

- TH1 - The threat of inadvertent damage to the equipment;
- TH 2 - The threat of improper shutdown of equipment;
- TH 3 - The threat of inadvertent deletion of files with important information;

Intentional subjective Information Security threats:.

- TH 4 - The threat of deliberate physical destruction of the system;
- TH 5 - The threat of the scrapping of the most important components of the Information System;
- TH 6 - The threat of shutdown subsystems provide Information System;
- TH 7 - The threat of the scrapping of the subsystems provide Information System;

Technogenic threats:

- TH8 - The threat of failure assistive technology;
- TH9 - The threat of failure of power supply system;
- TH10 - The threat of failure, the climate control system;

Vulnerability of information systems of the university. The main vulnerabilities of information systems in educational institutions are:

- V1 - Having unlocked built-in accounts
- V2 - incorrectly set access rights to information resources
- V3- The presence of unused potentially dangerous services and PP
- V4 - Incorrect configuration protection
- V5 - Low level of qualification of the IS staff
- V6 - Low levels of qualifications users
- V7 - Improperly organized access to hardware IP
- V8 - incorrectly implemented concurrent access to software
- V9 - incorrectly defined user rights
- V10 - Improperly organized storage media
- V11 - Improperly organized records of media
- V12 - Missing or improperly organized system of anti-virus

Controls of Information Security in University. Thus, the main measures and procedures of information security at the university are:

- C1 - Organization of procedures for the storage of documents
- C2 - Develop procedures for rapid response to incidents
- C3 - Administrative and technical means of monitoring the work of
- users
- C4 - Use of licensed certified
- C5 - Restriction of access to the software
- C6 - Technical support for hardware resources
- C7 - Backup
- C8 - Learn the basics of information security staff
- C9 - Corporate Culture
- C10 - Measures to prevent conflicts in the team
- C11 - The development of the internal regulatory documents for the IS

Therefore in expert system through the holding procedure of questionnaire that is described in paper [4] we define Threats (THi), Vulnerabilities (Vi) and Controls (Ci) rules of expert systems knowledge engine. The received rules are in the following:

- R1: IF V1 and V2 are LOW THEN TH5 is HIGH and C1 is LOW
- R2: IF V4 and V3 are LOW THEN TH5 and TH6 is HIGH and C2 is LOW
- R3: IF V7 is LOW THEN TH5 is HIGH and C1 is LOW
- R4: IF V4 and V6 are LOW THEN TH5 is HIGH and C1 is LOW
- R5: IF V1 and V2 are LOW THEN TH2 is HIGH
- R6: IF V11 is VERY LOW THEN TH4 is HIGH and C7 is LOW
- R7: IF V1 and V2 are LOW THEN TH6 is HIGH and C10 is VERY LOW
- R8: IF V1 and V2 are LOW THEN TH7 is HIGH and C11 is LOW

Thus, according to the rules above we a can calculate which controls are low and needed to be maintained again. This distinguished approach of ontology helps to organize and define appropriate rules according to ISO standards. And according to controls we can generate recommendations for this particular situation.

Summarizing sample above we can say that:

Information resources management of higher education in current conditions is impossible without scientific evidence and practical implementation of a balanced policy of information security [34-36].

Higher education institutions have a number of features that must be considered when building a system of information security. Specificity of information security in the educational institution is that it is - a public institution with volatile audience, as well as a place of increased activity "beginners cyber criminals."[35]. Features of the university as an object Information is also associated with a multi-character activities, the variety of forms and methods of educational work, the spatial distributed infrastructure (branches, representative offices), and so on.

5 Ontology and ISO 2700K Mapping

Since we are developing an expert system based on the standard of information security, there is a need to find conformity between the point of standards (ISO 2700k) and of the ontology that we developed above. Let us consider the table of ontology that was prepared by using help of experts in Information Security auditing process [36]

Based on the work [25], we can see that the standards are crossing in items with the ontology which is given above. Let's look at a few examples of where we can make a topology on table 2. Before we have to define vulnerability categories and threats in numbered way (see table1). It is a rough subcategory that is based on threats and vulnerabilities table.

Table 1. Identification of vulnerability and threat

ID of Vulnerability	Vulnerability
1	Subjective Vulnerability
2	Objective Vulnerability
3	Occasional Vulnerability
ID of Threat	**Threat**
1	Unintentional subjective Information Security threats
2	Intentional subjective Information Security threats
3	Technogenic threats
4	External Intentional Male Factor unauthorized access
5	External unintentional External legal user Authorized access
6	Internal Unpremeditated Staff errors

Table 2. Mapping of clauses and given vulnerabilities and threats

Clause	Vulnerability	Threat
Security Policy	3,2,1	6,2
Organization of Information security	2,1,3	4,6,5,2
Asset Management	1,2,3	6,3,4,2
Human Resource Security	1,2	6,3,2,4
Physical and Environmental Security	2,3,1	6,5,4,3,2
Communications and Operations Management	2,3,1	5,4,6,3,1
Access control	1,3,2	6,5,4,2,1,3
Business Continuity Management	2,3,1	6
Compliance	1	6

We can mention that the Internal Unpremeditated Staff Errors are common and popular threat, then External Intentional Male Factor unauthorized access threat is the second most popular factor. According to Vulnerability table Subjective Vulnerabilities are most common whereas Occasional Vulnerability is the second often meeting in this ISO standard.

6 Conclusion

Summing up all the above, we can state that the task of improving the security of information systems and technologies in current conditions characterized by complexity, uncertainty, related to the large number of internal and external factors affecting information security. To solve the problem of information security, particularly all you need to perform the identification of assets and set the initial level of security that meets in the Information Systems. The identification process should consider the main characteristics of asset: information value, the sensitivity of the assets to the threats, the availability of protective measures. It is necessary to note that among the factors that affect the safety, special place is occupied by the subjective factors that potentially are the most dangerous.

Ontology of knowledge base security of information systems can be described as following:

- problem area of complex security tasks
- information systems (concepts that define the essence of information security, and communication between them);
- threat to safety-critical Information Systems (external and internal objective factors, external and internal subjective factors);
- measures and technologies to provide a comprehensive security of information systems;
- a comprehensive methodology for Information Systems Security (heuristic knowledge about the state of security of information systems, strategies for integrated Security)
- identification of the assets of Information Systems, the definition of criteria and performance security, development assessment procedures of the criteria and indicators to develop a model integrated provision of Information Security, etc.
- principles of integrated Information security (systemic, adaptability, transparency and confidentiality, continuity, learning and the accumulation of experience, etc.).

Thus, the solution of the problem area provides complex security (the aggregate of the basic concepts that define the essence of the study, and the links between them), including: the subject area purpose of the study, tasks, possible tactics and strategies used to achieve this goal. The analysis in this study allowed us to construct an ontological scheme of subject area with in-depth branches and the main criteria for the definition of such concepts as vulnerability, threat, to classify the sources of threats and vulnerabilities Security of Information Systems, to identify the characteristics of the process of building models offender, to identify the main types of attacks on information resources identify the main types of damage to information assets, to classify the methods of information and build ontological model of the information security threats. In order to achieve these purposes we proposed development of Vulnerability&Threat Security ontology mapping to ISO Standards where we find out the most common vulnerabilities and threats.

We have considered an example of the use of information security ontology in practice example to check the vulnerability of the university. This example clearly shows the using and mapping of rules, types of threats and vulnerabilities, and controls. As we showed the operating with these concepts is very helpful in the process of building the knowledge base for expert system in the field of information security audit.

References

1. Val Thiagarajan, B.E.: BS 7799 Audit Checklist (2002), `http://www.sans.org/score/checklists/ISO_17799_checklist.pdf`
2. ISO IEC 27002 2005 Information Security Audit Tool(2005), `http://www.praxiom.com/iso-17799-audit.htm`
3. Stepanova, D., Parkin, S., Moorsel, A.: A knowledge Base For Justified Information Security Decision-Making. In: 4th International Conference on Software and Data Technologies (ICSOFT 2009), pp. 326–311 (2009)
4. Atymtayeva, L., Kozhakhmet, K., Bortsova, G., Inoue, A.: Methodology and Ontology of Expert System for Information Security Audit. In: Proceedings of the 6th International Conference on Soft Computing and Intelligent Systems and the 13th International Symposium on Advanced Intelligent Systems, Kobe, Japan, pp. 238–243 (2012)
5. Atymtayeva, L., Kozhakhmet, K., Bortsova, G.: Some Issues of Development of Intelligent System for Information Security Auditing. In: Proceedings of the International Conference of Computational Intelligence and Intelligent Systems 2012, London, UK, June 1-2, vol. 2, pp. 725–731 (2012)
6. Atymtayeva, L., Kozhakhmet, K., Bortsova, G., Inoue, A.: Expert System for Security Audit Using Fuzzy Logic. In: Proceedings of the 23rd Midwest Artificial Intelligence and Cognitive Science Conference, MAICS, Cincinnati, USA, April 21-22, pp. 146–151 (2012), `http://ceur-ws.org/Vol-841/`
7. Atymtayeva, L., Akzhalova, A., Kozhakhmet, K., Naizabayeva, L.: Development of Intelligent Systems for Information Security Auditing and Management:Review and Assumptions Analysis. In: Proceedings of the 5th International Conference on Application of Information and Communication Technologies, Baku, Azerbaijan, October 12-14, pp. 87–91 (2011)
8. Tsudik, G., Summers, R.: AudES - an Expert System for Security Auditing. IBM Los Angeles Scientific Center (1990)
9. Fenz, S., Ekelhart, A.: Formalizing information security knowledge. In: ASIACCS 2009: Proceedings of the 2009 ACM Symposium on Information, Computer and Communications Security. ACM (2009)
10. Threats catalogue on Information Systems Information technology — Security techniques — Code of practice for information security management (2005)
11. ISO/IEC 27002:2005, Information technology — Security techniques — Code of practice for information security management (2005)

12. http://www.odbv.org
13. Maljuk, A.A.: Information Security: Contemporary Issues. Security Information Tehnologiy (1), 5–9 (2010)
14. Domarev, V.V.: Safety of information technology. The System approach - Kiev, p. 992. Publishing House, Diasoft (2004)
15. Maljuk, A.: On the intensification of information security. Security of Information Technology (1), 6–10 (2011)
16. Maljuk, A.A.: Information security; conceptual and methodological framework for the protection of information. Textbook. allowance for vuzov. M: Hot
17. Line-Telecom, - 280c (2004)
18. Gerasimenko, V.A., Maljuk, A.A.: Framework for the protection of information. MEPI, Moscow (1997)
19. Maksimov, V.I., Kornoushenko, E.K.: The analytical basis for the use cognitive approach in solving semi-structured problems
20. Proceedings of the Institute of Control Sciences. - M.,- T. 2, pp. 95–109 (1999)
21. Castells, M.: The Information Age: Economy, Society and Culture. In: Shkaratan, O.I. (ed.) Per. from English. Under Teach., Moscow (2000)
22. Azhmuhamedov, I.M.: The principles of integrated security Information Systems. Journal ASTU. Series: Governance
23. Computer Engineering and Computer Science 1, 7–11 (2011)
24. Skorodumov BI on conceptual and terminological apparatus
25. Information Security. BIT (4), 43–45 (2008)
26. The Federal Law of 27.12.2002, 184-FZ (as amended. 1 December 2007). On technical regulation. Confident 2, C.44–C.49 (2002)
27. Nikolaev, Y.I.: Designing secure information technology, p. 312. St. Petersburg State Technical Publishing, St. Petersburg (1997)
28. Rumyantsev, O.G., Dodonov, V.N.: Legal encyclopedia, Moscow, vol. INFRA-M (1997)
29. Dotsenko, E.L.: Psychological Manipulation 295 (1996)
30. Crimean, L.A.: Organizational Data Protection, http://www.dsec.ru/about/articles/stuff
31. Vihorev, S.V.: The classification of information security threats
32. Cnews.ru annual review of Network attacks and system information Safety 2001, http://www.elvis.ru/informatorium_networks.shtml
33. Lukatckiy, A.: Detection of attacks. BHV-Petersburg, St. Petersburg (2003)
34. Vihorev, S.V., Kobtsev, R.Y.: How to identify the sources of threats? Open number of 7-8/2002g, http://www.elvis.ru/files/howto.pdf
35. http://art.thelib.ru/computers/security/mir_atak_mnogoobrazen.html # ixzz1PQX
36. http://bga.com.tr

Fuzzy Logic and Neural Network Based Induction Control in a Diesel Engine

B. Ganesh Kumar[1], S.R. Shankapal[1], A.S. Ravindran[2], Keith Burnham[3], and E. Ramakrishnan[4]

[1] M.S. Ramaiah School of Advanced Studies, Bangalore, India
ganesh.vinu@gmail.com
[2] Department of Mechanical Engineering, Alpha college of Engineering, Bangalore, India
[3] Faculty of Engineering and Computing, Coventry University, Coventry, United Kingdom
[4] Robert Bosch Engineering and Business Solutions Limited, Coimbatore, India

Abstract. To achieve the superior real time fuel economy, to meet the consistent stringent automotive exhaust emissions or to ensure best energy utilization, lots of new technologies are being adopted by the automotive manufacturers and suppliers. This new technologies comes with more and more complex to the existing system. This leads to the increase in the calibration parameters and indirectly affects the calibration efforts too.

In addition to this, due to the deterioration or failure of the engine components like, exhaust gas treatment devices, intake devices etc., are resulting in high emissions or unexpected uncomfortable driving for the drivers. To resolve this, there is a need for a flexible and intelligent control strategy.

Currently, the available conventional control strategies use the mapping method. The calibration time is long and the work is complex when adopting this mapping method. The model based control strategies also not successful in governing the unexpected behavior in the system.

Hence, a new controller based approach for an air system, based on hybrid of fuzzy logic and neural network is proposed in this research work to control the air mass, EGR (Exhaust Gas Recirculation) ratio, boost pressure and inter cooler. This new approach is designed to be implemented in a standard ECU (Electronic Control Unit) without any change in the current engine hardware design. The fuzzy logic based controller will replace the existing conventional map based PID controller. The neural network will learn the deterioration and failures of engine components and perform online calibration for the fuzzy logic controller.

Thus the combination of fuzzy and neural network approach will help to avoid the high emissions and unexpected uncomfortable driving mode for the drivers. The proposed, new control approach, which uses the hybrid approach of fuzzy logic and neural network, is very easy to tune, simplify the development time, improve the control precision of the air system and reduce cost and time of calibration.

Keywords: Fuzzy logic, neural network, air control, intelligent control.

Y.I. Cho et al. (eds.), *Soft Computing in Artificial Intelligence*,
Advances in Intelligent Systems and Computing 270,
DOI: 10.1007/978-3-319-05515-2_8, © Springer International Publishing Switzerland 2014

1 Introduction

To achieve reduced exhaust emissions imposed by the legislation such as Euro V, VI [1], superior fuel economy and improved engine performance led to lots of advanced technologies in the diesel engine systems.

Technologies such as exhaust gas recirculation (EGR), variable geometry turbo charging (VGT) and variable valve actuation system have played a key role in achieving expected requirements. However these advanced technologies come at the cost of increased control system complexity. The full potential of these devices are difficult to achieve due to the limitation in the convention control strategies [2]. In addition to that, aging of the components of the diesel engine also contribute emissions and bad performance [6]. These led to the need for the flexible, high performance oriented and enhanced control strategies.

Intelligent control strategies, such as fuzzy logic, neural network, genetic algorithms or combination of these have shown many advantages in engineering system control and modeling. They have the ability to model rapidly and learn the characteristics of multi–variant complex systems, exhibiting advantages in performance over more conventional mathematical techniques.

In this paper, a new control approach for an air system of diesel engine based on hybrid of fuzzy logic and neural network is proposed to control the air mass, exhaust gas recirculation ratio and boost pressure. This approach is designed to be implemented in ECU without any changes to current engine hardware design.

2 Engine Management Architecture

Architecture consists of elements and rules of how these elements interact to provide functional architecture. The term 'functional architecture' defines principal declarations of how many subsystems a system network is divided into, how these subsystems interact and which axioms these modular structures are subjected to. Only if such future functional architecture is agreed upon can modules of these sub-functions be re-used and alternative solutions exchanged for identical job steps.

The engine management software architecture is layered architecture. This is represented in the figure 1.

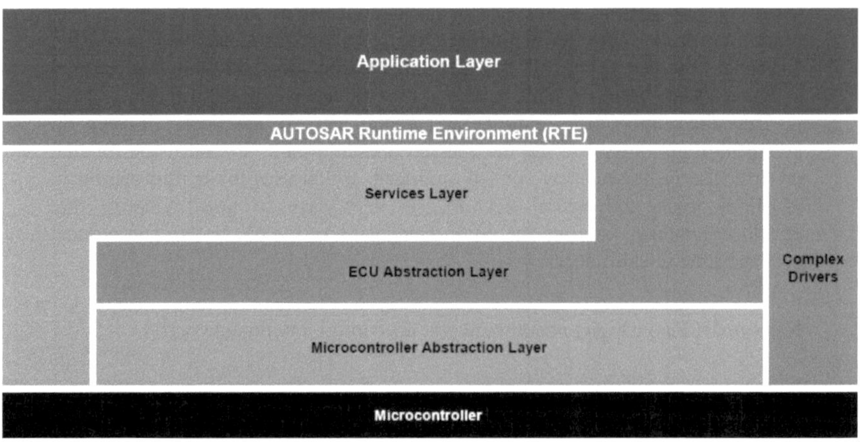

Fig. 1. Layered Architecture - Level 0

In the layered architecture, application layer is required to provide the engine related core functions like air system, exhaust system. The device encapsulation layer or ECU abstraction layer is required to support the different kind of input, output devices, actuators and sensors like throttle valve, exhaust gas recirculation valve, temperature sensors, etc.,. This represented in the figure 2 and 3.

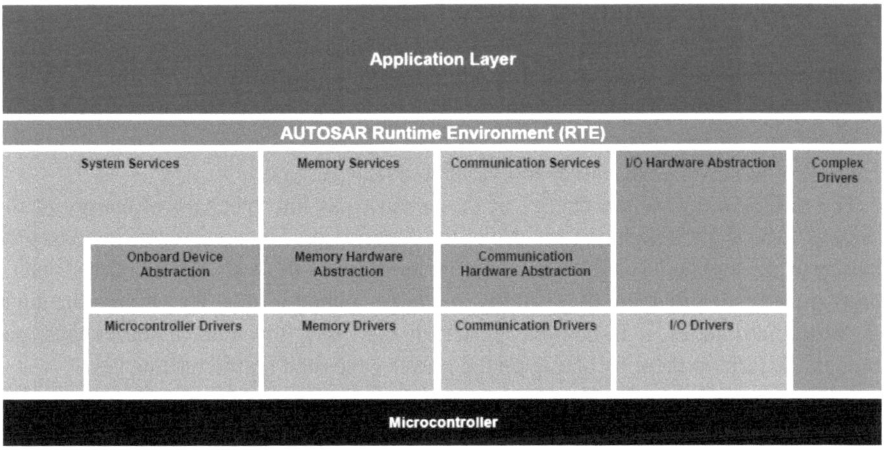

Fig. 2. Layered Architecture - Level 1

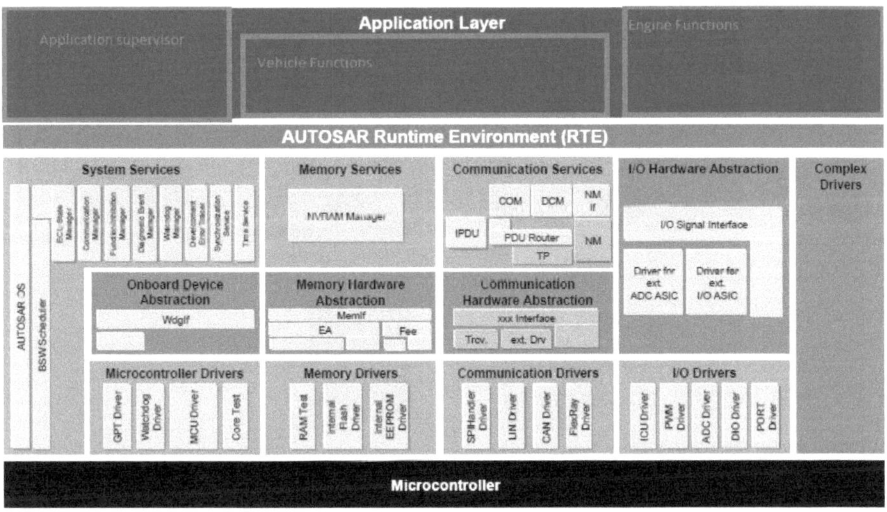

Fig. 3. Layered Architecture - Level 2

In this layered architecture, the application software consists of vehicle functions and engine functions whereas the device driver layer will get the required inputs from the upper layer known as application layer / application software.

The current proposed intelligent controller is implemented in application software (air mass (AIRCTL) and boost pressure (PCR) governor implementation) and device encapsulation layer (position governor implementation) in the layered engine management architecture.

3 Induction System of a Diesel Engine

In this research work, a diesel engine equipped with a variable nozzle turbine (VNT) and an exhaust gas recirculation system (EGR) is considered as shown in the figure 4. This system consists of air filter, sensors, EGR cooler bypass valve, charge air cooler, linear throttle valve, EGR valve, and variable nozzle turbine.

The turbine converts the energy of the exhaust gas into mechanical energy of the rotating turbo shaft, which, in turn, drives the compressor. The compressor increases the density of air supplied to the engine. The larger mass of fresh air can be burned with a larger quantity of fuel there by resulting in a larger output torque. By varying the angle of the turbine blades, it is possible to act on the mass flow rate of the exhaust gas through the turbine there by acting on the power generated by the turbine [2].

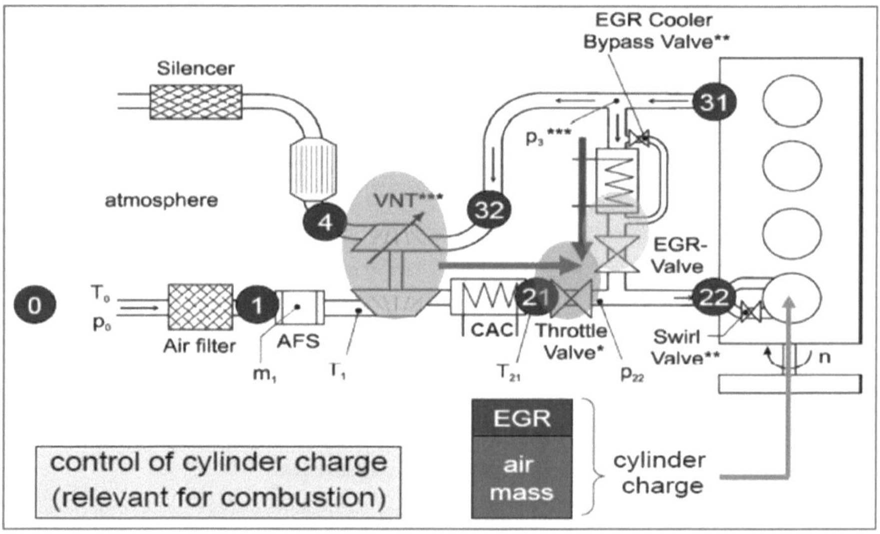

Fig. 4. Induction system of a diesel engine

To reduce the emissions of harmful nitrogen oxides (NOx) produced during combustion, a portion of the exhaust gas can be diverted back to the intake manifold to dilute the air supplied by the compressor. This process is referred to as exhaust gas recirculation (EGR). It is accomplished with an EGR valve that connects the intake manifold to the exhaust manifold. In the cylinders the recirculated exhaust gas acts as inert gas thus lowering the flame temperature and, hence decreasing the formation of NOx [2].

Our diesel engine induction system includes a throttle between the compressor and the intake manifold as well. This throttle permits the air system to create a variable pressure drop through the EGR valve, thereby increasing the EGR rates [2].

The engine configuration used in the experiments is mentioned below.

Table 1. Engine Configuration

Engine Configuration	
Number of cylinders	4
Swept Volume	2.21
Bore * Stroke	88 * 88.4 mm
Power	103 kw @ 3400 rpm
Maximum Torque	315 Nm @ 1600 – 2400 rpm
Aspiration	TCIC with cooled EGR and VGT

4 Conventional Control Strategy

In order to obtain the best result from the EGR system, the most important thing is to control the mass flow rate of the EGR accurately. The best results are calibrated as a tradeoff function as shown in the figure 5.

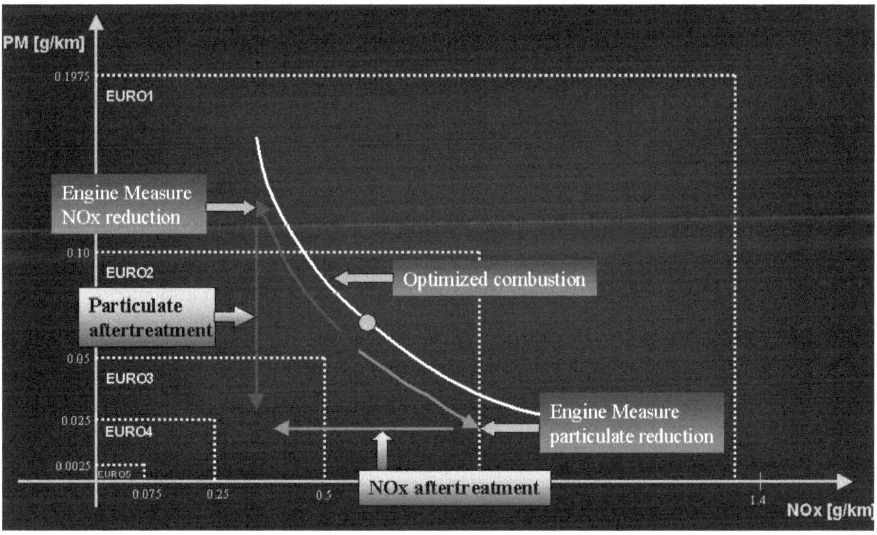

Fig. 5. Tradeoff curve for the calibration

The ECU (Engine Control Unit), send a duty cycle command to the EGR valve based on the engine operation condition. Then the EGR valve opens to allow the exact amount of exhaust gas recirculation for the current running condition [3].

The variable nozzle turbine can provide different values of the boost pressure for any engine condition by modifying the orientation of the turbine blades; hence the variable nozzle turbine regulates the power transmission from the rotor to the shaft [4]. The control action is accomplished by the ECU to which the error between the actual boost pressure and the target reference is fed.

The standard control strategy for VNT, EGR and throttle systems uses the proportional + integral + derivative control structure with a feed-forward term as shown in the figure 6.

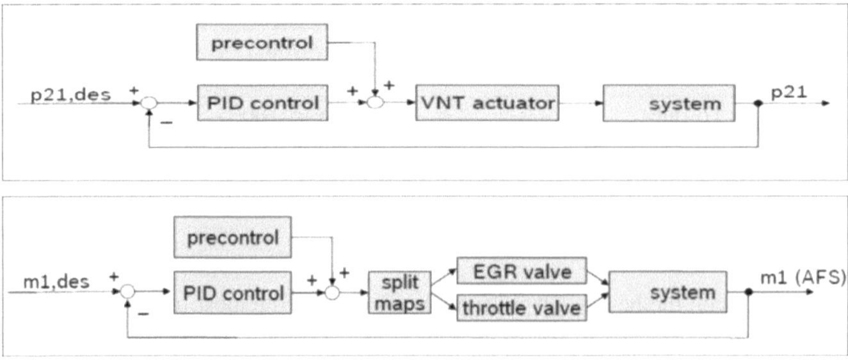

Fig. 6. Conventional control strategies

Classical control methods, such as Proportional, Integral and Derivative control, do not guarantee a satisfactory behavior at each operating point, due to high system non-linearity, aging of mechanical parts and environment conditions [4]. The open loop term can never guarantee accurate set point tracking due to the engine variability, aging and driving environment. The response of boost pressure and airflow to the VNT and EGR system varies with engine operating points; therefore gain scheduling is employed extensively [1].

Facing the future standard requirements, conventional strategy seems to have reached its limits and should be replaced by the intelligent control strategies.

5 Implementation of New Proposed Intelligent Control Strategy

To demonstrate the improvement in the controller performance of the system, reduction in calibration parameters and intelligent behavior of the new controller the following objectives formulated and executed step by step.

1. To review the performance of the conventional induction control system and establish the relationship between control system performance for emission and fuel consumption
2. To measure the performance of the existing induction control system and to obtain emission and fuel consumption characteristics for a baseline diesel engine
3. To develop fuzzy logic based control algorithms to enhance the performance of the conventional induction control system

4. To test and compare the result for control system performance against baseline system
5. To develop the learning algorithms based on neural networks for the developed fuzzy based control system to self- tune, online

As a result of objective 1, the diesel engine, induction system related parameters and its relationship between emissions and fuel consumptions are identified. They are shown as in the figure 7 and 8.

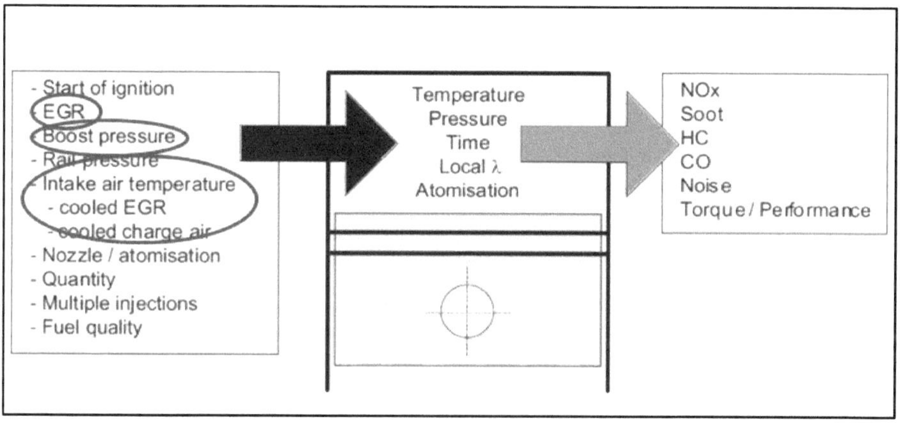

Fig. 7. Induction system parameters and its relationship with emissions - I

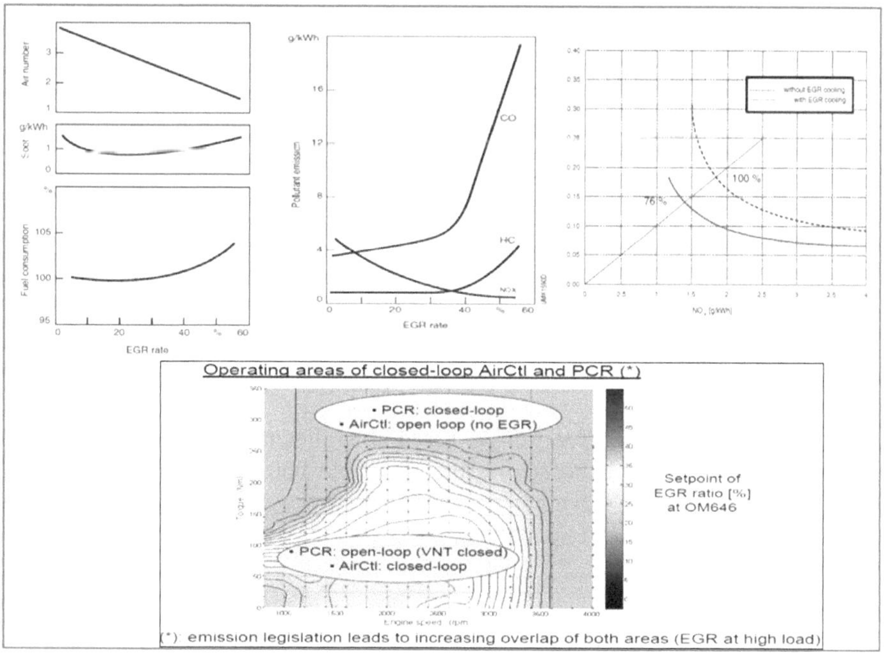

Fig. 8. Induction system parameters and its relationship with emissions – II

The objective 2 was carried out at the vehicle with the mentioned configuration (figure 4). The following results were obtained and they framed as a baseline value to compare against the new system as shown below.

Table 2. Baselined count of calibration parameters

Calibration parameters	
Air Control (EGR + throttle)	
Numbers of MAPs	115
Number of Curves	61
Number of parameters	257
Boost Pressure Control	
Numbers of MAPs	70
Number of Curves	21
Number of parameters	102
Actuator	
Number of MAPs/Actuator	10
Number of Curves/Actuator	20
Number of parameters/Actuator	250

Table 3. Baselined controller performance

Controller Performance Target	
Steady state error	< 1%
Setting time (Unit step 0-100%)	< 50ms
Rise time (Unit step 0-100%)	< 200ms
Overshoot	<= 2%

In the objective 3, new fuzzy logic controller was designed and implemented in the ECU. It was implemented with the help of ASCET, a model and simulation tool. A new index based algorithm was implemented to ensure efficient and quick search of required rule. This newly designed fuzzy logic controller is implemented along with the conventional control system with a selectable switch. This is show in the figure 9 and 10.

The newly implemented fuzzy logic based controller uses the require air mass, available air mass, require boost pressure, available boost pressure, current VNT actuator position, throttle valve position, EGR valve position, engine speed, engine temperature as inputs and required controller air mass, boost pressure, actuator positions as outputs.

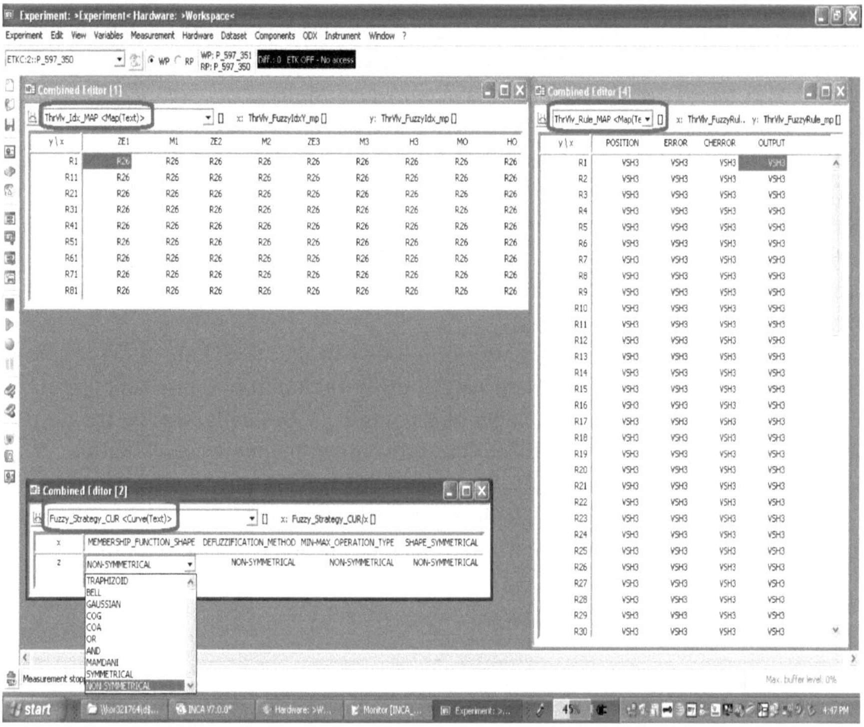

Fig. 9. New fuzzy rule indexing algorithm

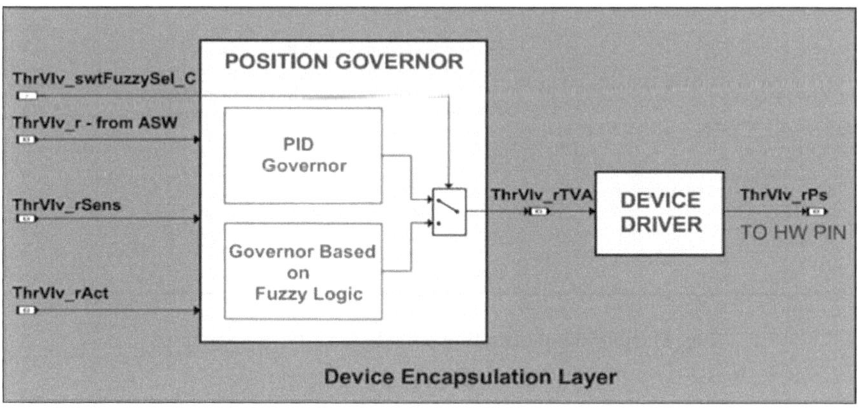

Fig. 10. Fuzzy logic based position governor

The symmetrical triangular membership's functions were used for the inputs and outputs for fuzzification. Centre of gravity method is used a defuzzification method. Min and Max based fuzzy interference system was designed.

As per the objective 4, the newly designed and developed fuzzy logic based control system was tested in the same vehicle. The results are explained in the next section.

Considering the difficulties in demonstrating the objective 5, an engine model was developed in GT-power environment. The developed engine model was validated and verified against the vehicle data used for objective 4. The fuzzy logic controller also developed in MATLab and validated and verified against the vehicle data used for the objective 4.

A neural network based model was developed in the MATLab to determine the required air mass and boost pressure values using the trade of curve function during the aging or failure of the components of the engine. Basically, during the aging or failure of the components or based on preventive maintenance situation of the components, the operating point of the system will be calibrated online by the neural network. This calibrated value will be input to the designed fuzzy logic controller. Thus the neuro fuzzy logic controller tunes itself intelligently according to the situation.

They are shown in below figures.

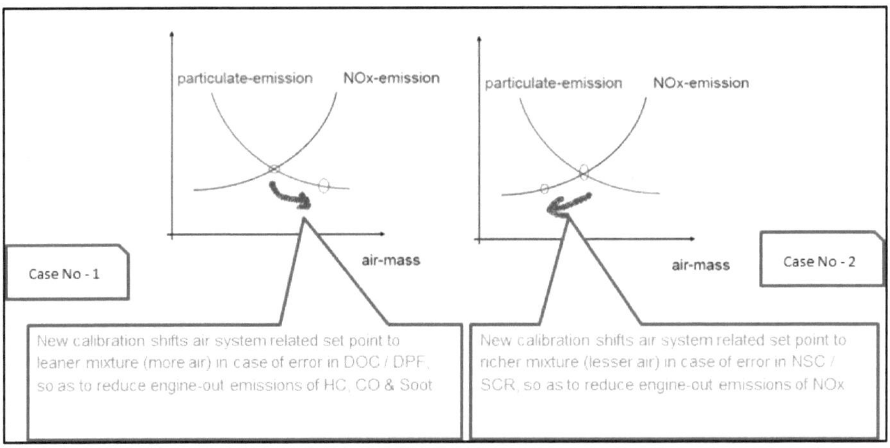

Fig. 11. Shifting of operating point based different scenarios

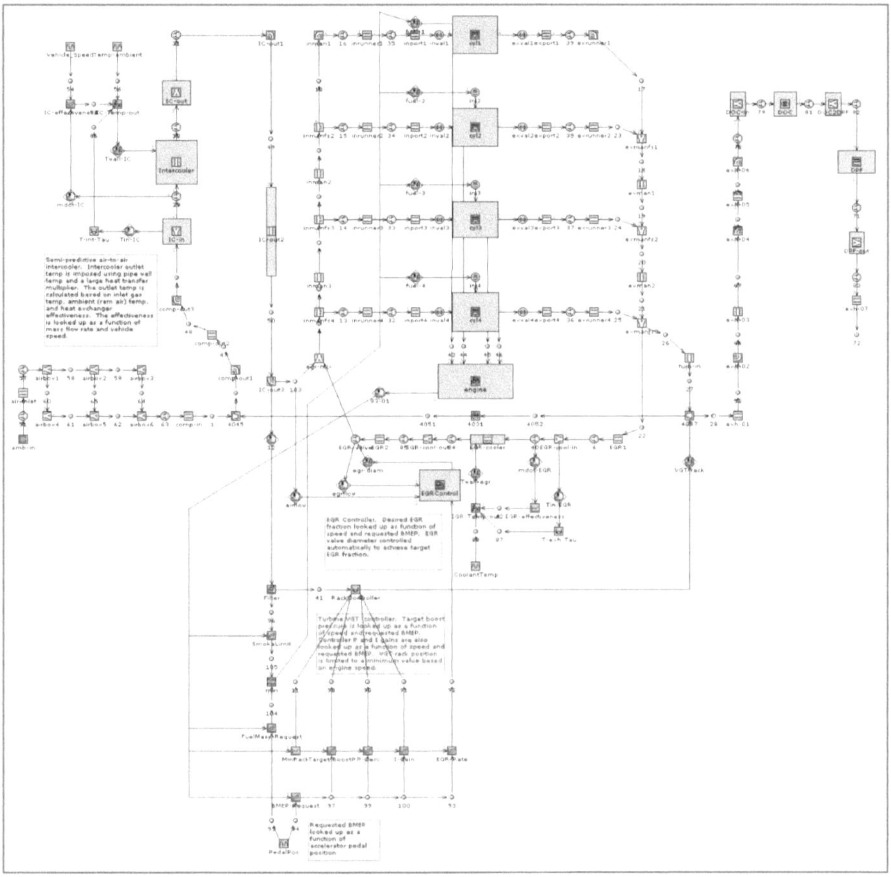

Fig. 12. GT Power based engine model

6 Test Results and Discussions

The fuzzy logic controller performance was improved compared to the conventional system.

- Compared to the conventional PID control the fuzzy logic control system performance was improved and less oscillation. These are depicted in the below figures.

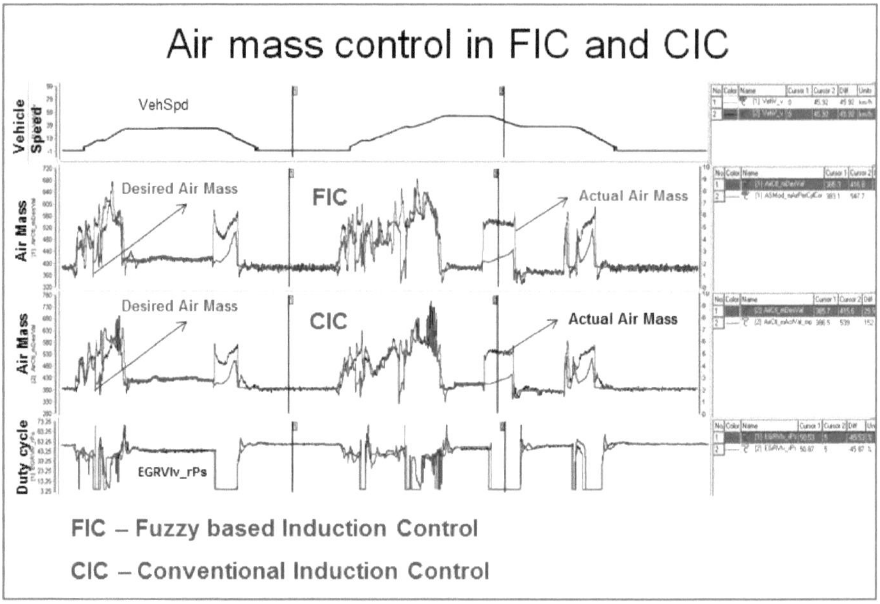

Fig. 13. Comparison of fuzzy logic control and conventional control for fresh air mass

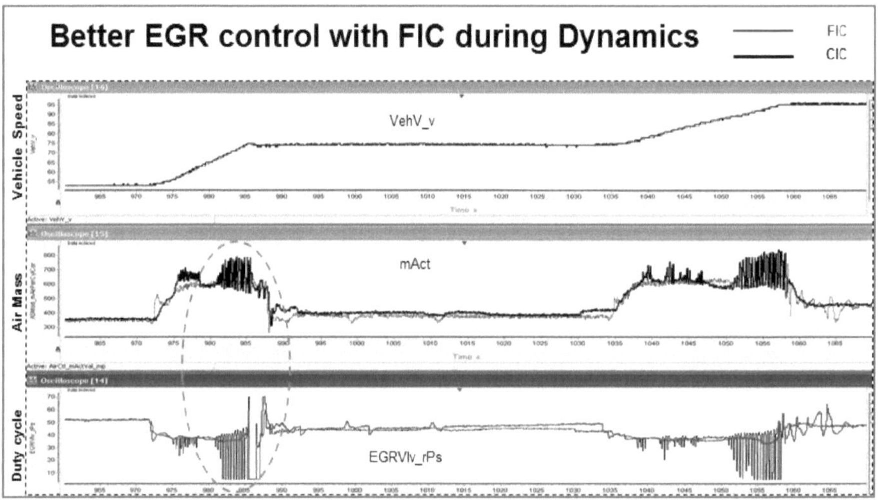

Fig. 14. Comparison of fuzzy logic control and conventional control for EGR control

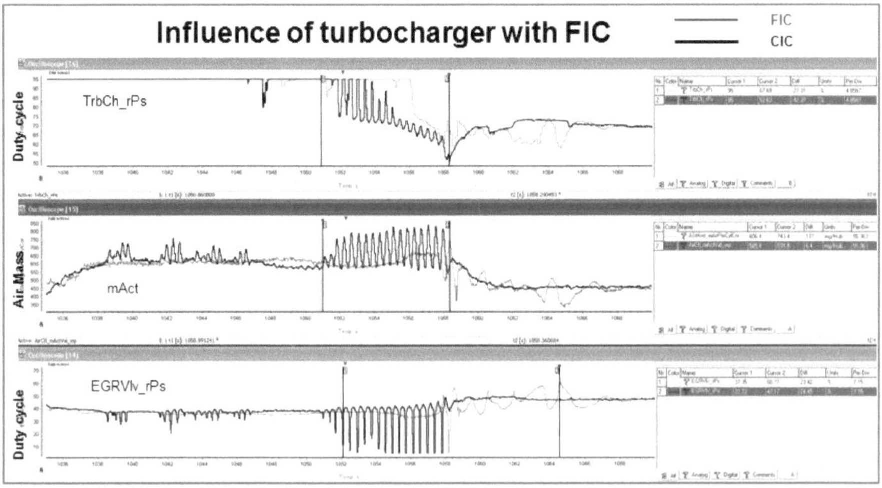

Fig. 15. Comparison of fuzzy logic control and conventional control for PCR control

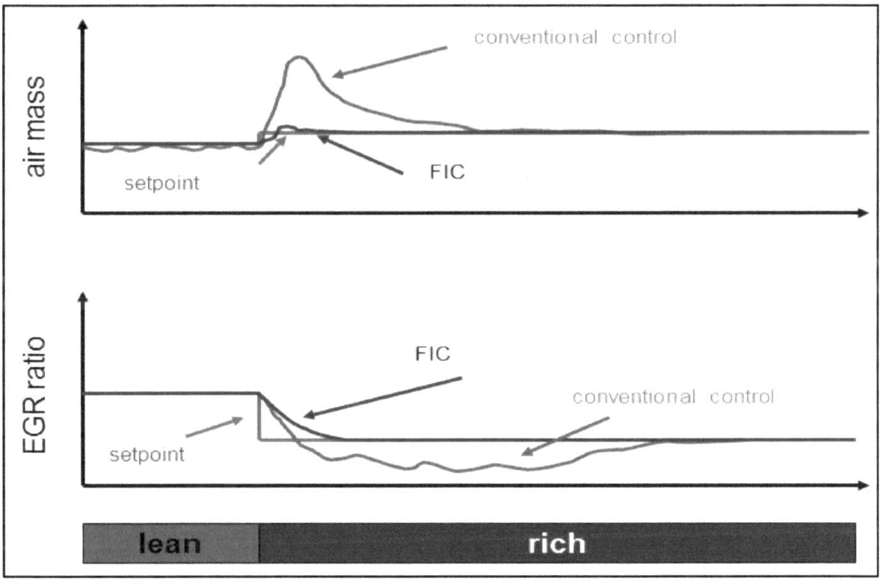

Fig. 16. Comparison of fuzzy logic control and conventional control for induction control

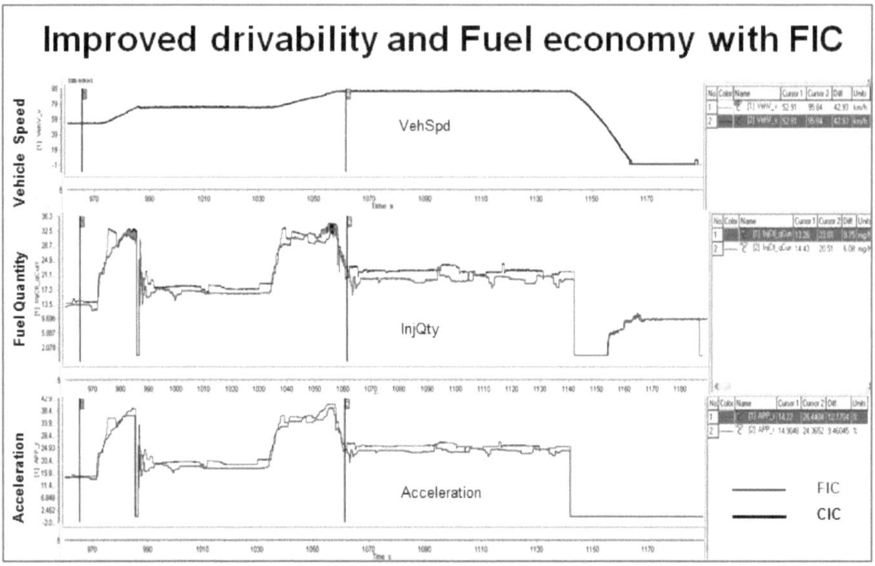

Fig. 17. Comparison of fuzzy logic control and conventional control for drivability and fuel economy control

- Comparatively there was around 50% reduction in the calibration parameters used for the air control, boost pressure control and actuator control.
- The determination of aging and fault mechanism of the components were accurately done by the neural network. These were simulated by the changing the GT-Power components sizes and preventive maintenance parameters.
- The shifting of operating points was very successful to control the emissions in control.

7 Conclusions

In this experimental study, the fuzzy logic and neural network based has been applied comparatively to a diesel engine induction control to demonstrate

(a) Improvement in the controller performance of fuzzy logic against PID control
(b) Reduction of calibration parameters by using of fuzzy logic control
(c) Intelligent decision and control of neural-fuzzy logic control.

The biggest advantage of fuzzy logic compared to conventional method is easier implementation, easier calibration. The combination of fuzzy logic with neural network gives improvement to tune the fuzzy systems.

In this study, the demonstration of intelligent decision was performed for only one scenario. This can improve by using different scenarios and different engine.

In future, the same study can be combined with the "big data and cloud based technology" to do the best online calibration.

References

1. Arnold, J.F., Langlosis, N., Chafouk, H., Tremouliere, G.: Control of the air system of a diesel engine: A fuzzy multivariable approach. In: Proceedings of the 2006 IEEE International Conference on Control Applications, Munich, Germany, October 4-6 (2006)
2. Wijetunge, R.S., Brace, C.J., Hawley, J.G., Vaughan, N.D.: Fuzzy Logic Control of Diesel Engine Turbo charging and Exhaust Gas Recirculation. University of Bath, UK
3. Cui, H.: Exhaust Gas Recirculation Control in a Spark-Ignition LPG Engine Using Neural Networks. In: Proceedings of the 6th World Congress on Intelligent Control and Automation, Dalian, China, pp. 21–23 (2006)
4. Dotoli, M., Lino, P.: Fuzzy Adaptive control of a variable geometry turbocharged diesel engine
5. Liu, B., Huang, M., Yang, X., Xia, X.: The State Key Laboratory of Advanced Design and Manufacturing for Vehicle Body. In: An Optimization of EGR Control System for Gasoline Using Fuzzy PID Strategy. Hunan University, Changsha
6. Obodeh, O., Ajuwa, C.I.: Mechanical engineering department, Ambrose Alli University, Ekpoma, Aigeria, Calibration of Aging Diesel Engine with Artificial Neural Networks. European Journal of Scientific Research 24(4), 520–531 (2008) ISSN 1450-216X
7. Atkinson, C.M., Long, T.W., Hanzevack, E.L.: Virtual sensing: A neural network based intelligent performance and emissions prediction system for on-board diagnostics and engine control. In: International Congress and Exposition, Detroit, Michigan, February 23-26. SAE Technical paper series (1998)
8. Tasdemir, S., Saritas, I., Ciniviz, M., Allahverdi, N.: Artificial neural network and fuzzy expert system comparision for prediction of performance and emission parameters on a gasoline engine. Expert Systems with Applications 38, 13912–13923 (2011)

Evaluation and Analysis of Relationship between Roles and Individual's Role Adaptation in Team Sports Using SOM

Data Analysis and Introduction of Theory

Yasunori Takemura, Muneyuki Yokoyama, Sho Omori, and Ryo Shimosaka

Nippon Bunri University, Faculty of Eng., Dept. of Mechanical and Electrical Eng., Japan
takemura@nbu.ac.jp

Abstract. Victory or defeat in team sports depends on each player's technique, physical strength, and psychological condition. It follows that team performance depends on the player's adaptation to (suitability for) a certain role (position in the team) and the relationships between different roles. We assume that team performance is related to physical and psychological features. Many researchers have proposed that physical features determine a player's suitability for a position. Psychological features have also been researched as factors of position adaptation. However, each feature has been investigated independently. The present research aims to develop a clustering method that considers both physical and psychological features in judging an individual's role and adaptation in the game. This paper reports the concept of the algorithm and result of psychological data analysis using self-organizing maps and principal component analysis.

Keywords: clustering, data mining, self-organizing map, principal component analysis.

1 Introduction

Victory or defeat in team sports depends on each player's individual technique, physical strength, and psychological condition. Similarly, it can be said that the suitability of an individual to a certain position in the team affects the team's performance.

In previous research, Barry and Cureton [1], Nicks and Fleishman [2], Larson [3], McCloy and Young [4] and others clustered physical features and conducted factor analysis in investigating sports performance. In Japan, Tokunaga studied the diagnosis criteria for athletic adaptation (i.e., suitability) in sports [5]. These works showed that physical features are one of the strongest factors determining athletic adaptation. However, Matsuda [6] showed that an athlete, no matter how good his/her physical features, is not athletically suited to team sports without having good motivation in terms of setting goals and training. That is to say, for an individual or team to be successful, a player needs to have both good physical features (e.g., techniques,

Y.I. Cho et al. (eds.), *Soft Computing in Artificial Intelligence*,
Advances in Intelligent Systems and Computing 270,
DOI: 10.1007/978-3-319-05515-2_9, © Springer International Publishing Switzerland 2014

balance, height, and weight) and good psychological features. For example, Saijo [7] presented the psychological features of Japanese and New Zealand rugby players.

In this way, the suitability of a player in a certain position and the relationships between different positions in team sports are related to physical and psychological features. As it stands now, a coach or selector decides the player suitability and relationships between positions him/herself. However, does it follow that good decisions are made? Previous research has not clarified athlete adaptation to positions and relationships between positions considering both physical and psychological features.

In the present study, we develop a clustering algorithm for positioning adaptation and relationships in team sports. We consider the two main types of features, which are physical and psychological features, and introduce the concept of using an algorithm to cluster player features in team sports. Additionally, we present input data and an analysis of the input data using conventional statistical methods (e.g., principal component analysis).

2 Research Plan and Method

2.1 Data and Choice of Sport

We apply a team-sport clustering algorithm to rugby. Rugby is selected because a rugby team has a large variety of positions and player attributes. A rugby game is called a match. It is a competition between two teams. Each match lasts for 80 minutes plus time added to account for injuries and stoppages during the match. A match consists of two halves of 40 minutes each. It is controlled by a single referee on the field of play, and two assistant referees (touch judges). The team with the greater number of points at the end of the match wins. Normally, a rugby team has a maximum of 15 players on the field, and seven substitutes. Each player has a specialized position. There are eight forwards, numbered 1 to 8, and seven backs, numbered 9 to 15. Player number 1 is a prop (PR), number 2 is the hooker, number 3 is another prop (PR), numbers 4 and 5 are locks (LOs), number 6 is the blindside flanker (FL), number 7 is the open-side flanker, number 8 is the "number 8" (No. 8), number 9 is the scrum-half (SH), number 10 is the fly-half (or stand-off, SO), number 11 is the left wing (WTB), number 12 is the inside center (CTB), number 13 is the outside center (CTB), number 14 is the right wing (WTB) and number 15 is the fullback (FB).

The main role of the forwards is to gain and retain possession of the ball. They take part in set pieces of the scrum and line-out. Generally, forwards are larger than backs, and are thus stronger but slower. Forwards also play a role in taking the ball forward, and generally do so by driving into the opposing forwards. Back-row forwards such as flankers and the number 8 are becoming increasingly athletic and fast, staying out of the breakdown (where possession of the ball is contested) to participate in running moves with the backs.

The role of the backs is to gain territorial advantage and score points by running with the ball or kicking the ball. The SH will gain possession of the ball from the forwards and usually feed it to the SO, who then controls how the attacking team will proceed. The backline will tend to score tries (where the ball is grounded behind the opposing team's try (goal) line) through the tactical placement of players, creating holes in the opposition defense line. A successful backline movement will cause the opposition defense to commit to many players at strategic points, thus opening up space for the faster, outside backs (the wings and fullback) to run into [8].

To provide a method of determining player adaptation to certain roles and the relationships between those roles, we first obtain and analyze the physical and psychological features of rugby players using a conventional statistical method.

2.2 Physical Features and Analysis

Physical data of members of the Nippon Bunri University (NBU) rugby club were recorded in 2010. The NBU rugby club in 2010 comprised 56 students (14 fourth-grade students, 14 third-grade students, 12 second-grade students and 16 first-grade students). There were 32 forwards (FWs) and 24 backs (BKs).

Physical features recorded were height [cm], weight [kg], body fat [%], neck length [cm], brachium length [cm] (R: right and L: left), chest circumstance [cm], waist circumstance [cm], hip circumstance [cm], length of thigh (R, L) [cm], length of calf (R, L) [cm], anteflexion while standing (flexibility) [negative value, cm], number of abdominal crunches completed in 30 s, bench-press weight [kg], number of squats completed in 30 s, number of chin-ups completed in one effort, and the distance run in 7 minutes [m]. The data set thus had 19 physical dimensions. Some members played more than one position. There were 12 PRs, 10 LOs, 12 FLs, six No. 8s, six SHs, five SOs, 10 CTBs, seven WTBs and eight FBs.

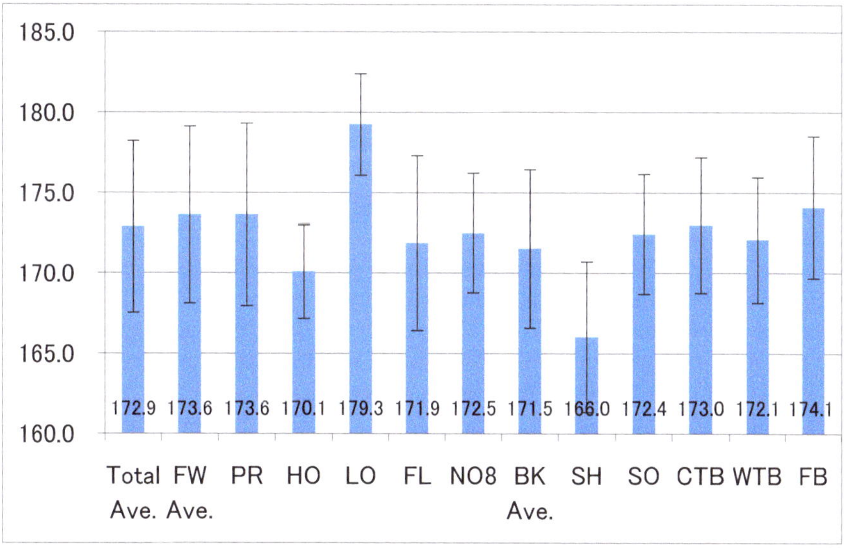

Fig. 1. Heights of rugby players (NBU rugby club in 2010)

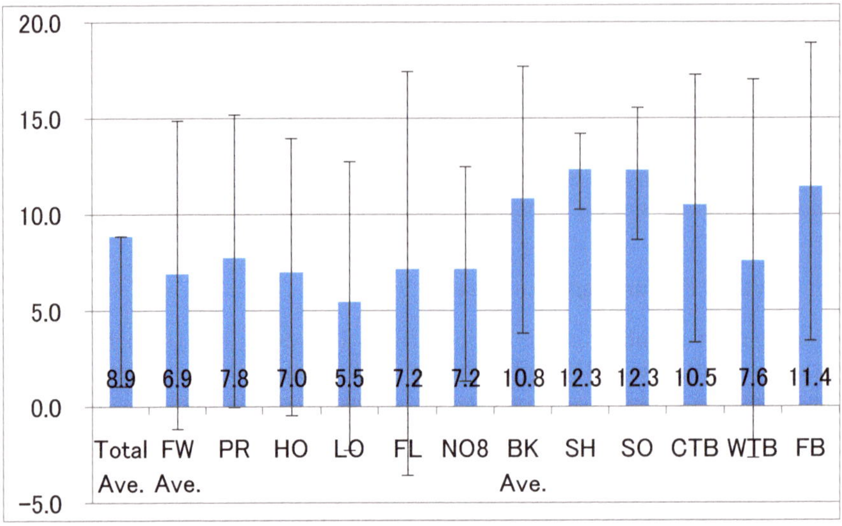

Fig. 2. eflexion in standing (flexibility) of rugby players (NBU 2010th)

Figure 1 shows the height data. The horizontal axis denotes individual positions, the FW group, the BK group and the entire group. The vertical axis shows the height [cm]. The average height is greatest for LOs, and least for SHs. The LO plays a role in securely locking the scrum (which is a set piece). The job of the LO, who is in the second row of the scrum, is to push firmly on the first row of the scrum. Additionally, the LO is a designated jumper in the line-out, which is a set play for restarting the game once the ball has gone out of play over a sideline. Therefore, the most important feature of the LO is to be tall. In contrast, the SH handles the ball as it comes out of a scrum (set piece), maul (contesting of the carried ball in open play) or ruck (contesting of the ball on the ground in open play). At the moment that the SH receives the ball from the dense sea of FWs, the SH is rushed by opposing FWs. The FW thus requires agility and judgment in kicking, passing or running the ball. They therefore do not need to be tall, and since they regularly gather the ball and pass from the ground, there is an advantage in being short.

We are unable to find distinctive trends for all physical features. For example, Fig. 2 shows the result of flexibility; no trend is observed in the data. The horizontal axis denotes the individual positions, the FW group, the BK group and the entire group. The vertical axis shows vertical flexibility while standing (anteflexion) [cm]. The data vary greatly for each position. The figure suggests that flexibility is not related to a player's suitability to a position. Thus, trends are revealed for some physical features but not others.

2.3 Psycological Features and Analysys

Psychological data of members of the NBU rugby club were recorded in 2011. The NBU rugby club in 2011 comprised 48 students (12 fourth-grade students, 10 third-grade students, 12 second-grade students and 14 first-grade students).

Psychological features measured were those of the Diagnostic Inventory of Psychological Competitive Ability for Athletes (DIPCA.3) [9]. DIPCA.3 measures 12 types of psychological ability in a 48-item questionnaire. DIPCA.3 is often used before mental training, because it reveals athletic strengths and weaknesses. The DIPCA.3 check sheet consists of 48 questions that measure psychological ability and four questions that measure reliability. These questions have already been analyzed by good–poor analysis (G-P analysis), and the answers provide 12 criteria relating to five factors. The factors are motivation in sport, mental stability and concentration, confidence, operation capability and cooperativeness. Additionally, we measure the reliability of the answers by comparing answers to similar questions. To examine these factors in detail, 12 criteria are desciribed. Motivation in sport consists of four criteria: endurance, fighting spirit, self-realization motivation and motivation to win. Mental stability and concentration consist of three criteria: the ability to relax, capacity to concentrate and self-control. Confidence consists of two criteria: determination and confidence. Operation capability consists of two criteria: predictive capability and judgment. Cooperativeness has only one criterion, which is simply described as cooperativeness. DIPCA.3 provides a total score (ability). Each criterion is scored, and the score of each factor is the sum of scores for the related criteria. The total score is calculated by summing the score for each factor. Table 1 gives the total scores, which are classified into five levels. A classification of 1 indicates a very low overall mental ability and a classification of 5 indicates a very high ability. Each mass describes the total score. The methodology of DIPCA.3 is such that each person answers the 52 questions on a scale of 1 to 5 (1: I don't think so, 2: I rarely think so, 3: I sometimes think so, 4: I often think so, 5: I always think so) within 10 to 15 minutes.

To analyze the team psychological condition, we use principal component analysis (PCA) [10]. The central idea of PCA is to reduce the dimensionality of a data set comprising a large number of interrelated variables, while retaining as much of the variation present in the data set as possible. In this case, 12 criteria are used in the data set. PCA is conducted by constructing a new set of variables, the principal components (PCs), which are uncorrelated and ordered so that the first few retain most of the variation present in all of the original variables. In this experiment, we used 48 sets of data for 12 criteria (i.e., one data set has 12 dimensions).

Table 2 shows the eigenvalue distribution for the 2011 data sets. If we use principal components 1 and 2, the cumulative contribution ratio is almost 65%. Therefore, we check the factor loading of principal component 1, and decide that principal component 1 denotes total mental ability. Additionally, we check the loading factor of principal component 2, and find that the positive direction denotes motivation in sport (especially a motivation to win), while the negative direction denotes mental stability and concentration.

Figure 3 is a scatter plot of principal components that also shows the total score. The horizontal axis denotes principal component 1 (total points), and the vertical axis denotes principal component 2 (motivation in sport to mental stability and concentration).

Table 1. Total Score of DIPCA.3

Judgment	1 (very poor)	2 (poor)	3 (a little poor)	4 (good)	5 (excellent)
Man	<141	142–164	165–186	187–209	>210
Woman	<131	132–154	155–178	179–202	>203

Table 2. PCA Evaluation value

Principal component	Eigenvalue	Contribution ratio	Cumulative contribution ratio
1	5.443	45.36%	45.36%
2	2.350	19.58%	64.94%
3	1.591	13.26%	78.20%
4	0.622	5.18%	83.38%
5	0.532	4.43%	87.82%
6	0.389	3.24%	91.06%
7	0.296	2.46%	93.52%
8	0.257	2.14%	95.67%
9	0.188	1.57%	97.24%
10	0.139	1.16%	98.40%
11	0.128	1.07%	99.47%
12	0.064	0.53%	100.00%

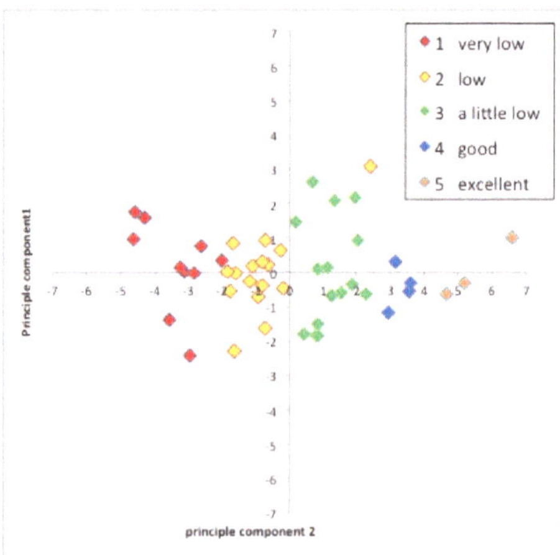

Fig. 3. Scatter plot of the principal component result showing the total score

Each plotted point gives principal components 1 and 2. Additionally, each plotted point is colored according to the player's total score. As the eigenvalue of principal component 2 increases, we find that the player's total score increases. Therefore, the horizontal axis indicates the total score.

Figure 4 is a scatter plot of principal components showing the motivation to win. The horizontal and vertical axes are the same as those of Fig. 3. Each plotted point describes principal components 1 and 2. Additionally, each plotted point is colored according to the result of the motivation to win. We find that positive eigenvalues of principal component 1 correspond to high motivation to win. Figure 5 is a scatter plot showing mental stability and concentration. The horizontal and vertical axes are the same as those of Figs. 3 and 4. Each point denotes principal components 1 and 2, and

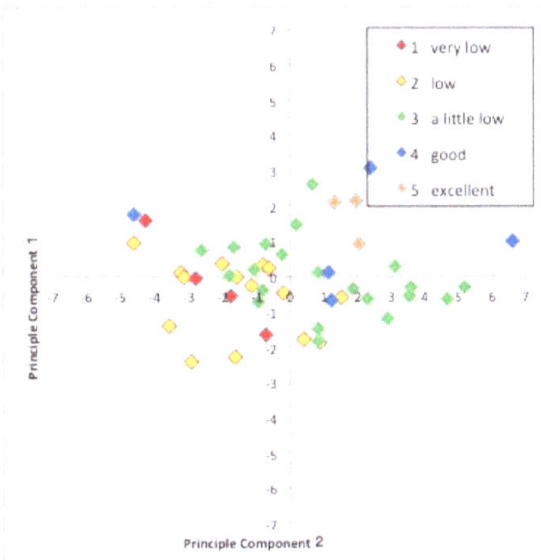

Fig. 4. Scatter plot of the principal component result showing motivation to win

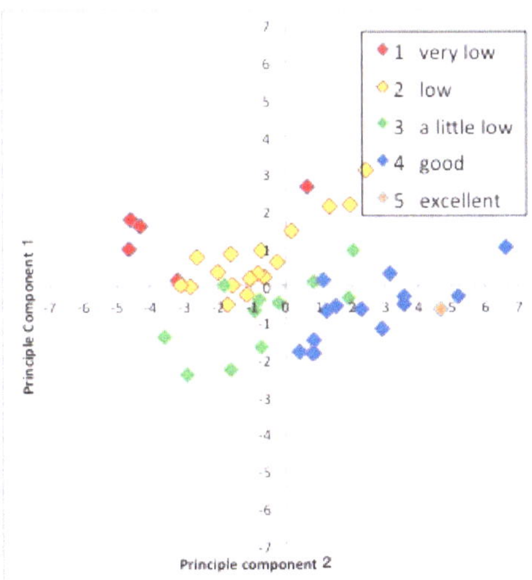

Fig. 5. Scatter plot of the principal component result showing mental stability and concentration

is colored according to the result of mental stability and concentration. Negative eigenvalues of principal component 1 correspond to high mental stability and concentration.

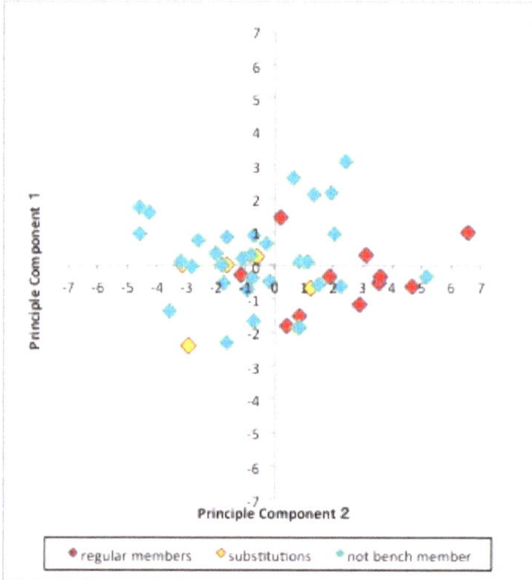

Fig. 6. Scatter plot of the principal component result showing whether individuals were regular or occasional players

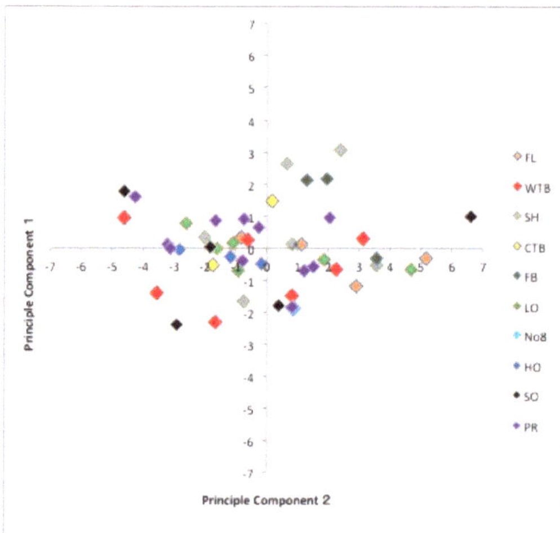

Fig. 7. Scatter plot of the principal component result showing position

Figure 6 is a scatter plot showing whether players were regular players, regular substitutes or occasional players. The plot suggests that regular players were selected by the coach for their total mental score. That is to say, most substitutes and occasional players require mental training. Figure 7 is a scatter plot showing each position; no clear trend is observed in this case.

3 Proposed Method (Using Self-organizing Map)

The analysis of physical and psychological features indicates that the conventional statistical method cannot reveal player adaption to positions in team sports and relationships between positions. We therefore propose a new method using self-organizing maps (SOMs). An SOM is a topologically correct feature map proposed by Kohonen and is well known as an attractive tool for extracting the characteristics of data and classifying data into clusters through its self-organizing process. The brains of many higher-order animals appear to achieve a topological relationship through a stream of sensory inputs, and several algorithms have been suggested to duplicate this neural processing. The SOM has an algorithm that is capable of establishing a feature map by learning a random sequence of input samples. Kohonen's algorithm can be represented in a simple iterative form, thereby demonstrating its computational power. The SOM is an unsupervised learning algorithm for generating a topological transformation from a high-dimensional data vector space to a low-dimensional (usual one-dimensional or two-dimensional) space. Higher-dimensional maps are also possible, but are not discussed here. The neurons are programmed for various input patterns or classes of input patterns through a competitive learning process [11, 12]. The SOM algorithm includes four processes: an evaluation process, competitive process, cooperative process and adaptive process.

Table 3. list the variables used in a SOM algorithm

Parameter	Description
θ	Training data
x	Input vector (physical features) [j = 1–19]
y	Input vector (psychological features) [j' = 1, ... 12] (12 criterion scores)
i	Index denoting class (i = 1, /..., l)
w	Reference vector
k	Index expressing unit (k = 1, ..., K)
E	Distance between input vector and reference vector
k*	Best matching unit (BMU)
ϕ	Neighbor function
$d(a,b)$	Euclidean distance between a and b
ψ	Learning rate
σ	Neighbor radius
τ	Time constant

Table 3 lists the variables used in a SOM algorithm. The training data are defined by Eq. (3.1), with suffix i denoting the input data set. x and y are normalized from -1 to 1 for input into the SOM algorithm.

$$\theta_i = (x_j, y_{j'}) \tag{3.1}$$

1. Evaluation Process

In the evaluation process, θ_i is input into all units, and the square mean errors (differences) E_i^k between θ_i and reference vector w_k are calculated. The upper suffix k is the number of units and i is the index of the data set.

$$E_i^k = \left\| w^k - \theta^k \right\|^2 \tag{3.2}$$

2. Competitive Process

In the competitive process, to find the best matching input vector θ for reference vector w, the best matching unit (BMU) (k^*) is selected in Eq. (3.3). That is, the closest environment to θ is selected in the w space.

$$k_i^* = \arg_k \min E_i^k \tag{3.3}$$

3. Cooperative Process

The learning quantity for each module is calculated according to Eq. (3.6). The learning distributional rate ψ_i^k expresses the quantity to be distributed to the k-th unit with regard to the i-th data class. Here, the neighborhood function $\phi(.)$, like a Gaussian function, decreases monotonically with an increase in the distance function $d(.)$, and the iteration of learning t is selected.

Let $d(k, k_i^*)$ denote the Euclidean distance between the k-th unit and k_i^*, which is the BMU for the i-th data class. The amplitude of the topological neighborhood decreases monotonically with increasing distance $d(k, k_i^*)$.

$$\phi_i^k = \exp\left(\frac{d(k, k_i^*)^2}{2\sigma^2} \right) \tag{3.4}$$

The parameter σ is the effective width of the topological neighborhood, and is called the neighbor radius. The use of σ allows the SOM algorithm to converge more quickly than a rectangular topological neighborhood function [13,14]. The neighbor radius σ (Eq. (3.5)) of the topological neighborhood function decreases with time.

$$\sigma = \sigma_{\min} + (\sigma_{\max} - \sigma_{\min})\exp\left(\frac{-t}{\tau}\right) \tag{3.5}$$

In a batch-type SOM, the learning rate of each unit is defined by ψ_i^k, which is normalized by the sum of the ϕ_i^k (Eq. 3.6).

$$\psi_i^k = \frac{\phi_i^k}{\sum_{i'} \phi_{i'}^k} \tag{3.6}$$

4. Adaptive Process
5. In the Kohonens SOM, all unit vectors are adjusted using Eq. (3.7) in the adaptive process.

$$w(t+1) = w(t) + \psi_i^k(\theta_i - w^k) \tag{3.7}$$

For the adaptive function, Eq. (3.8) is introduced instead of Eq. (3.7) since Eq. (3.8) is known to achieve faster convergence for a batch-type SOM.

$$w(t+1) = \sum_i \psi_i^k \theta_i \tag{3.8}$$

The SOM algorithm generates feature maps in two dimensions following processes (1)–(4). Each lattice cell describes a unit.

4 Evaluation and Discussion

To evaluate the performance of the SOM algorithm, individually, at first, only x input data sets (only physical data sets) are used. Figure 8 shows the feature maps used for the 2010 physical data sets. Each lattice cell describes a unit. If a unit is the BMU for a data set, it is labeled as such at top left. The color of the lattice indicates the Euclidian distance to the neighboring unit. For example, blue indicates that the neighboring unit is nearby, red indicates that it is far away, and green indicates that it is at mid-distance.

In Fig. 8, FWs gather at the top side and BKs below. Thus the physical data sets totally relate to position. For example, left top side gather the PR position members. On the other hand, right below side gather the SH and WHB position members. That's means, PR position and SH or WHB positions have opposite characteristics in physical features. However, looking at the example of the BMU for the second column of the seventh row (FLs and No. 8s), the BMU located on the lower side. However, FLs and No.8s is not BK positions. That is to say, this player may be well suited to playing as a BK. In this way, if we use this feature maps, we can find new possibility for some players. Additionally, the feature map reveals the neighboring BMU. Each BMU player has the same physical features. This result can be used to partner players in practice or make substitutes using players who are out of position but physically similar.

Next, the 48 psychological data sets for players in 2010 are separately input into the SOM algorithm. Figure 9 is a feature map of psychological data sets (only using *y* [12:criterion data] input data sets). Thus, if a lattice cell has a BMU, it is labeled according to the total points (1 to 5) (Total points are not the input of the SOM.). Players who have total score low are gathered at top left, while players scoring high are gathered at bottom right. And also, bottom right side almost gather the regular team members. On the other hand, left top side gather almost un-regular team members. In the point of the position, in the fig.9 cannot find groups. That's means, psychological data sets is not related as position, on the other hand, we found as follows : if the coach choice the regular members, psychological data is very important for selecting.

As a result, we can cluster psychological and physical data individually in SOM feature maps. Additionally, we reveal that the SOM is able to cluster features of players in team sports using physical or psychological data sets.

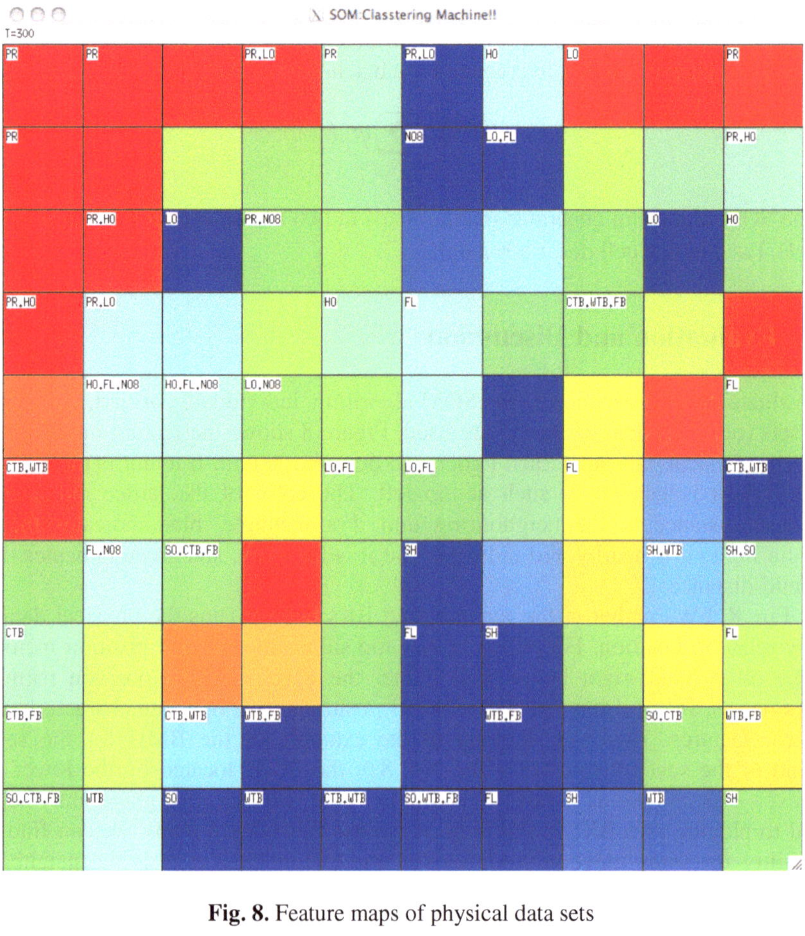

Fig. 8. Feature maps of physical data sets

Fig. 9. Feature maps of psychological data sets

5 Conclusion

In this paper, we proposed and developed an algorithm for clustering positions in team sports using an SOM to provide insights into player adaptation to positions and relationships between positions. Focusing on the data sets, conventional statistical analysis can find player adaptation to a position using physical data. However, if data are available for only a few physical and (especially) psychological features, player adaptation and the positional relationships cannot be revealed.

However, the SOM algorithm can reveal the positional adaption and relationships using high-dimensional data. In this paper, we individually input physical or psychological data sets into the SOM, and found trends for the team. In future work, considering the implementation of real team sport clustering, we need to combine physical features and psychological features for input into the SOM algorithm. As a result, we will show that the algorithm can cluster adaptation in team sports using physical and psychological data sets. In future work, we will discuss the result when inputting higher-dimensional data into the SOM algorithm.

Acknowledgment. This work was supported by a Grant-in-Aid for Young Scientists (B) 24700684 from JSPS KAKENHI.

References

1. Barry, A.J., Cureton, T.K.: Factor Analysis of Physique and Performance. Res. Quart. 23(3), 100–108 (1961)
2. Nick, D.C., Fleishman, E.A.: What Do Physical Adaptation Test Meassure? A Review of Factor Analysis Studies, Educational and Psycohological Measurement 22(1), 77–95 (1962)
3. Larson, L.A.: A Factor Analysis of Motor Ability Bariables and Tests for College Men. Res. Quart. 12, 499–517 (1941)
4. Mcloy, C.H., Young, R.D.: Test and Meassurement in Health and Physical Education, pp. 51–65. Applenton-Century-Crofts, Inc., New York (1954)
5. Tokunaga, M.: Factor Analysis about sports adaptation. Japan Journal of Physical Education, Health and Sport Sciences 22(2), 71–80 (1977)
6. Matsuda, I., Singer, R.N.: Motor learning and human performance. Taishukan-shoten (1986) (in Japanese) ISBN978-4-469-26119-6
7. Saijo, O., Suda, K., et al.: On the Psychologicl Aptitude of Rugby Foot Ball Players –From Comarision New Zealand University Selected Team with Nippon College of Physical Education Team –. Journal of Nippon College of Physical Education 22(2), 135–138 (1993)
8. A rugby rules and positions, `http://www.Rubyunionrules.com/`
9. Tokunaga, M.: Evaluation scales for athletes' psychological competitive ability: development and systematization of the scales. Japan Journal of Physical Education, Health and Sport Sciences 46(1), 1–17 (2001)
10. Jolliffe, I.T.: Principal component analysis, pp. 1–60. Springer ISBN0-387-95442-2
11. Kohonen, T.: Self-organized formation of topologically correct feature maps. Biol. Cybernetics 43, 59–69 (1982)
12. Kohonen, T.: The self-organizing map. In: Proceedings of the Institute of Electrical and Electronics Engineers, pp. 1460–1480
13. Lo, Z.-P., Fujita, M., Bavarian, B.: Analysis of neighborhood interaction in Kohonen neural networks. In: 6th International Parallel Processing Symposium Proceedings, Los Almitos, CA, pp. 247–249 (1991)
14. Lo, Z.-P., Yu, Y., Bavarian, B.: Analysis of the convergence properties of topology preserving neural networks. IEEE Transactions on Neural Networks 4, 207–220 (1993)

Design of Radial Basis Function Classifier Based on Polynomial Neural Networks

Tae Chon Ahn[1], Seok Beom Roh[1], Zi Long Yin[1], and Yong Soo Kim[2,*]

[1] Dept. of Electronics Convergence Engineering, Wonkwang University, Chonbuk Korea
{nado,tcahn}@wku.ac.kr, yinzilong1@hotmail.com
[2] Dept. of Computer Engineering, Daejeon University, Daejeon, Korea
kystj@dju.kr

Abstract. In this paper, to improve the generalization ability of radial basis function networks, we apply the polynomial neural networks as the virtual input variables of radial basis function networks. The parameters of each polynomial neuron are estimated by linear discriminant analysis. In each layer of polynomial neural networks, the polynomial neurons are selected in terms of the objective function of linear discriminant analysis.

Keywords: Linear Discriminant Analysis, Radial Basis Function Networks, Polynomial Neural Networks, Polynomial Neuron.

1 Introduction

Radial Basis Function (RBF) networks, which is well known to be able to deal with high dimensional problem, have been widely studied for various tasks in the fields such as regression and classification problem [1, 2, 3]. Especially, the structural topologies of these networks and learning methods have been the interesting issues among the researchers. In this paper, we define the radial basis function using Fuzzy C-Means Clustering [4] not the bell type function such as Gaussian function. Fuzzy sets are useful in solving real world problems when we encounter imprecise and uncertain information. In particular, the discriminant surfaces generated by RBF networks for classification problem seem to be a relatively simple. The simple discriminant surface comes from the linear combination of the radial basis functions in radial basis function networks. Some researchers proposed the polynomial based RBF NNs [5]. In the polynomial based RBF NNs, the outcome is the weighted summation of several polynomials. To enhance the variability of the networks, we use Polynomial Neural Networks (PNNs) [6, 7, 8]. In this paper, we propose a new approach referred to as Radial Basis Function classifier based on Polynomial Neural Networks where the virtual input variables of the consequent function are used as the output variables generated in polynomial neural networks. We demonstrate how to implement PNNs dynamically and to use PNNs as virtual input variables. To show

[*] Corresponding author.

Y.I. Cho et al. (eds.), *Soft Computing in Artificial Intelligence*,
Advances in Intelligent Systems and Computing 270,
DOI: 10.1007/978-3-319-05515-2_10,

the classification of the proposed classifier, we make experiments with using several machine learning data sets and compare the generalization ability of the proposed classifier with the well-known classifiers such as linear discriminant analysis (LDA) and support vector machine (SVM) [9].

2 Polynomial Neural Networks for Virtual Input Variables

Considering a representation of a highly nonlinear real system using a dynamically formed model, there are many open questions on how to choose meaningful variables among all available input and output variables and how to select a structure of the model itself. A Group Method of Data Handling (GMDH) introduced by A. G. Ivakhnenko [10] comes as one of the alternatives that helps alleviate the problem. GMDH is a self-organizing method, which is similar to an evolutionary algorithm based on the natural law of the survival of the fittest. While the generic GMDH is a convenient optimization method dealing with the structure of the model, it comes with a certain shortcoming. In the "conventional" GMDH, where two-input quadratic regression is being used as the basic polynomial, we encounter difficulties to represent the complex nonlinear system. In order to alleviate this shortcoming, self-organizing Polynomial Neural Networks (PNN) [6, 7, 8, 11] were proposed by Oh et al. The structure of PNN is similar to that of a feed forward neural network. PNN is not a statically organized network whose topology is predefined and left unchanged. To the contrary, we encounter dynamically generated networks whose topologies could be adjusted during the design process.

The PNN algorithm is based on the GMDH method and utilizes some classes of polynomials such as linear, modified quadratic, and cubic, etc. (refer to Table 1). By choosing the most significant input variables and an order of the polynomial among the various types of forms being available, we can construct the best polynomial neuron (PN) at each layer. Additional layers are generated until the best performance of the extended model has been achieved. This type of methodology leads to an optimal PNN structure.

Table 1. Different types of the regression polynomial forming Polynomial Neural Networks

Polynomial order	Structure of Polynomial
Type I : Bilinear polynomial	$a_0 + a_1 z_p + a_2 z_q$
Type II : Modified Bi-quadratic polynomial	$a_0 + a_1 z_p + a_2 z_q + a_3 z_p z_q$
Type III : Bi-quadratic polynomial	$a_0 + a_1 z_p + a_2 z_q + a_3 z_p z_q + a_4 z_p^2 + a_5 z_q^2$

Consider that the input-output data are given in the form

$$(\boldsymbol{x}_j;\ y_j) = (x_{1j}, x_{2j}, \cdots, x_{mj};\ y_j), \qquad j = 1, 2, \cdots, N$$

The input-output relationship of the above data captured by PNN algorithm can be described in the following manner.

$$\hat{y} = \hat{f}(x_1, x_2, \cdots, x_m) = c_0 + \sum_{k=1}^{m} c_k x_k + \sum_{k=1}^{m}\sum_{l=1}^{m} c_{kl} x_k x_l + \sum_{j=1}^{m}\sum_{k=1}^{m}\sum_{l=1}^{m} c_{jkl} x_j x_k x_l + \cdots$$

Where c_k denotes the k-th coefficient of the model. An overall architecture of the PNN is shown Figure 1.

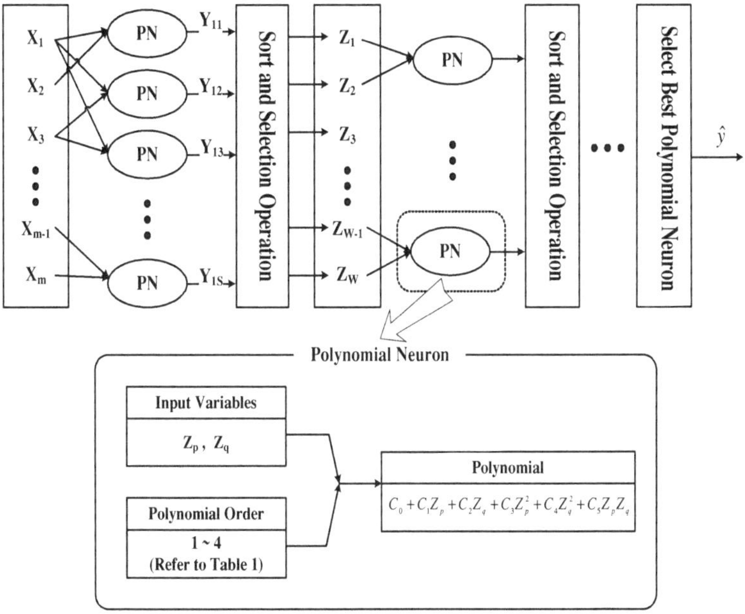

Fig. 1. An overall structure of PNNs

At the beginning of the optimization process to determine the output $\hat{y}$, we construct a PN form for each pair of independent variables according to the predefined number of the input variables available on a PN. Here one determines the parameters of the PN by invoking the least square method and using some training data. In this way, we choose the optimal model forming the 1st layer. In the sequel, we construct new PNs using intermediate variables (for example, z_m) being generated at the current iteration. Afterwards, we take another pair of new input variables, and repeat construction of PNs until a certain stopping criterion has been satisfied. Once the final layer has been constructed, the node characterized by the best performance is selected as the output node. All remaining nodes in that layer are discarded. Furthermore, all the nodes of previous layers that do not have influence the estimated output node are also removed. This removal is completed by tracing the data flow occurring at each iteration. The essence of the design is such that simple functions are combined at all nodes of each layer of the PNN, which leads to the realization of more complex dependencies. The outputs obtained at each of the nodes of the same layer are combined to produce higher order polynomials. The pseudo code describing the formation of the polynomial neural networks is shown as followings.

Main Procedure of Polynomial Neural Networks

1. Decide upon the design parameters such as
   (a) K: the number of input variables entering each node
   (b) M: total number of candidates (which can be chosen as the input variables at the next layer)
   (c) L: The number of layers of polynomial neural networks
2. Repeat
   (a) Call sub-procedure of polynomial neuron
   (b) Sort the values of the performance index obtained from the procedure of Polynomial Neuron in a descending order
   (c) Choose M candidates and set the selected M nodes as new input variables for the next layer
3. Until the predefined maximum number of layer L has been reached

Sub-Procedure of Polynomial Neuron

1. Repeat
   (a) Determine the structure of the polynomial (the selected input variables and the order of the polynomial)
   (b) Estimate the coefficients of a polynomial by using Linear Discriminant Analysis as (6)
   (c) Calculate the values of the performance index as (2)
2. Until $S = \frac{m!}{(m-k)! \cdot K!} \cdot T$ (T=4, m is the total number of input variables) has been reached
3. Return Sorted Values of the performance index

Fig. 2. Pseudo code for the construction of Polynomial Neural Networks

The parameters of each polynomial neuron are estimated by Linear Discriminant Analysis (LDA).

The j-th polynomial output Z_j is calculated as (1).

$$Z_j = \boldsymbol{W}^T \boldsymbol{X} \tag{1}$$

The objective function of LDA is defined as (2).

$$J = \frac{\boldsymbol{W}^T \boldsymbol{S_B} \boldsymbol{W}}{\boldsymbol{W}^T \boldsymbol{S_W} \boldsymbol{W}} \tag{2}$$

Where, S_B is the between class scatter matrix and is defined as (3) and S_W denotes the within class scatter matrix and is calculated by (4).

$$\boldsymbol{S_B} = (\boldsymbol{m_1} - \boldsymbol{m_2})(\boldsymbol{m_1} - \boldsymbol{m_2})^T,\ \ \boldsymbol{m_i} = \frac{1}{n_i}\sum_{x \in \boldsymbol{D_i}} \boldsymbol{X} \tag{3}$$

Here, D_i denotes the set of data which is involved in i-th class and n_i is the number of the data involved in D_i.

$$\boldsymbol{S_i} = \sum_{x \in \boldsymbol{D_i}} (\boldsymbol{X} - \boldsymbol{m_i})(\boldsymbol{X} - \boldsymbol{m_i})^T \tag{4}$$

$$\boldsymbol{S_w} = \boldsymbol{S_1} + \boldsymbol{S_2} \tag{5}$$

We calculate the value $\boldsymbol{W}$ which maximizes the objective function as (6).

$$\boldsymbol{W} = (\boldsymbol{S_W})^{-1}(\boldsymbol{m_1} - \boldsymbol{m_2}) \tag{6}$$

3 Radial Basis Function Networks Based on Polynomial Neural Networks

For the proposed classifier, the virtual input variables (which are not the input variables of the original system) are introduced to improve the approximation and prediction capabilities of the proposed classifier. As mentioned before, the advantage of the Polynomial Neural Networks is that in order to estimate values of numerous coefficients of a highly complicated polynomial by using the values of coefficients at each layer which have already been derived in an iterative fashion.

3.1 Radial Basis Function Networks

It is well known that the generic RBF NNs exhibits some advantages including global optimal approximation and classification capabilities as well as rapid convergence of the underlying learning procedures, see [12, 13, 14, 15].The generic topology of RBF NNs is depicted in Figure 3.

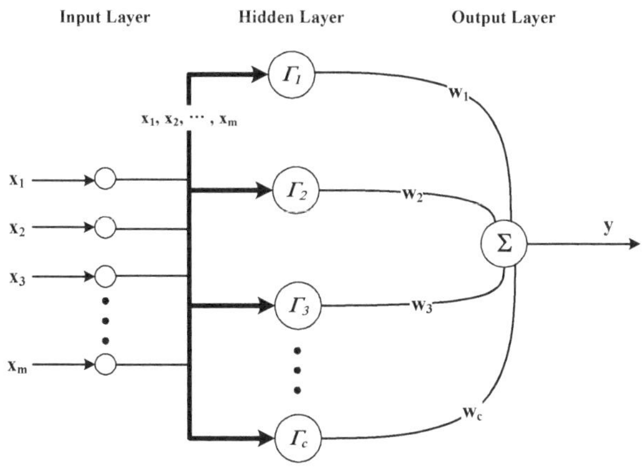

Fig. 3. General Architecture of Generic RBF Networks

The output of the generic RBF NN comes as a linear combination of the outputs($\Gamma(\mathrm{x})$) of the corresponding nodes at the hidden layer with the connection weights $w_1, w_2, \cdots, w_c$ as shown below

$$\hat{y}(\mathbf{x}) = \sum_{i=1}^{c} w_i \cdot \Gamma_i(\mathbf{x}) \tag{7}$$

Where $\Gamma_i(\mathbf{x})$ is the activation level of the i-th node present at the hidden layer.

The prototypes of the receptive fields (i.e., the nodes of the hidden layer) are determined by running fuzzy clustering. The output of each node in the hidden layer is an activation level of the corresponding linguistic term (fuzzy set).

$$\Gamma_i(\mathbf{x}) = \frac{1}{\sum_{j=1}^{c}\left(\frac{\|\mathrm{x}-\mathrm{v}_i\|}{\|\mathrm{x}-\mathrm{v}_j\|}\right)^2} \tag{8}$$

3.2 Radial Basis Function Networks Based on Polynomial Neural Networks

Unlike the generic RBF networks, the proposed RBF networks classifier has the extended type of the connection (weights) between the hidden layer and output layer. In the proposed RBF networks, we use polynomial neural networks rather than confining ourselves to some fixed numeric values.

The architecture of the proposed RBF networks is shown in Figure 4.

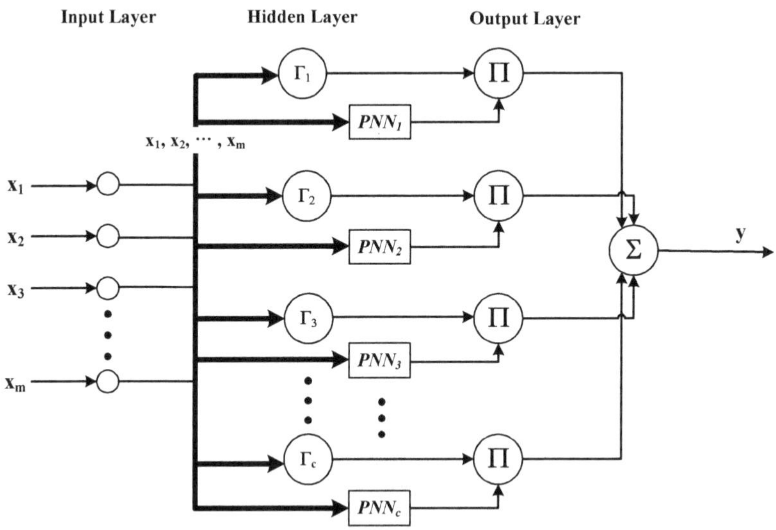

Fig. 4. Architecture of the proposed RBF networks based on PNNs

4 Experimental Study

In order to evaluate and quantify the classification effectiveness of the proposed classifier, the proposed classifier is experimented with by making use of a series of numeric data such as two synthetic datasets and several Machine Learning datasets (http://www.ics.uci.edu/~mlearn/MLRepository.html). In the assessment of the performance of the classifiers, we use the error rate of the resulting classifier.

We investigate and report the results of each experiment in terms of the mean and the standard deviation of the performance index.

We consider some predefined values of the parameters of the network whose values are summarized in Table 2. The choice of these particular numeric values has been motivated by the need to come up with a possibility to investigate of the performance of the classifier in a fairly comprehensive range of scenarios.

Table 2. Selected numeric values of the parameters of the proposed classifier

Parameter	Value
Polynomial Order (O)	1 (constant), 2 (linear) 3 (quadratic)
Number of input variables entering each node (K)	2 or 3
Total number of candidates (M)	10
Number of Layers (L)	3 or 5
Number of RBFs (c)	2, 3, or 5

Table 3 shows the comparison between the proposed classifier and the classification methods. In this experiments, we use 10 fold cross validation to evaluate the classification abilities and the final correct classification ratio is given in terms of its average and the standard deviation.

Table 3. Results of comparative analysis with the other classification method

	Australian	German	Heart	Ionosphere	Liver	PIMA
Proposed Classifier	***87.25±3.79***	***76.0±2.98***	***84.81±6.40***	***93.35±2.99***	***73.67±7.3***	***77.47±4.28***
LVQ3[9]	68.9	71.3	66.0	N/A	66.4	74.0
SVM[9]	65.0	70.0	60.8	N/A	68.1	75.7
SMO[16]	76.63	N/A	N/A	N/A	N/A	N/A
LBDA[17]	N/A	N/A	N/A	91.38±0.07	N/A	76.16±0.06

5 Conclusions

In this paper, we proposed the new design methodology of radial basis function networks based on Polynomial Neural Networks. The generalization ability of the proposed classifier is improved by using the output of polynomial neural networks as the input variables of radial basis function networks. The learning algorithm used to in the development of the conclusion part of the rules takes advantage of the linear discriminant analysis.

Acknowledgements. This paper was supported by Wonkwang University in 2013.

References

1. Renjifo, C., Barsic, D., Carmen, C., Norman, K., Peacock, G.S.: Improving radial basis function kernel classification through incremental learning and automatic parameter selection. Neurocomputing 72, 3–14 (2008)
2. Rocha, M., Cortez, P., Neves, J.: Simultaneous evolution of neural network topologies and weights for classification and regression. In: Cabestany, J., Prieto, A.G., Sandoval, F. (eds.) IWANN 2005. LNCS, vol. 3512, pp. 59–66. Springer, Heidelberg (2005)
3. Pedrycz, W., Park, H.S., Oh, S.K.: A granular-oriented development of functional radial basis function neural networks. Neurocomputing 72, 420–435 (2008)
4. Pedrycz, W.: Conditional Fuzzy C-Means. Pattern Recognition Letter 17(6), 625–632 (1996)
5. Oh, S.K., Kim, W.D., Pedrycz, W., Park, B.J.: Polynomial-based radial basis function neural networks (P-RBF NNs) realized with the aid of particle swarm optimization. Fuzzy Sets and Systems 163, 54–77 (2011)
6. Ivahnenko, A.G.: Polynomial theory of complex systems. IEEE Trans. on Systems, Man and Cybernetics SMC-12, 364–378 (1971)
7. Oh, S.-K., Pedrycz, W.: The design of self-organizing Polynomial Neural Networks. Information Science 141, 237–258 (2002)
8. Oh, S.-K., Pedrycz, W., Park, B.-J.: Polynomial Neural Networks Architecture: Analysis and Design. Computers and Electrical Engineering 29(6), 703–725 (2003)
9. Parades, R., Vidal, E.: Learning prototypes and distance: A prototype reduction technique based on nearest neighbor error minimization. Pattern Recognition 39, 180–188 (2006)
10. Farrow, S.J.: The GMDH algorithm. In: Farrow, S.J. (ed.) Self-organizing Methods in Modeling: GMDH Type Algorithms, Marcel Dekker, New York (1984)
11. Kleinsteuber, S., Sepehri, N.: A polynomial network modeling approach to a class of large-scale hydraulic systems. Computers Elect. Eng. 22, 151–168 (1996)
12. Er, M.J., Wu, S.Q., Lu, J.W., Toh, H.L.: Face recognition with radical basis function (RBF) neural networks. IEEE Transactions on Neural Networks 13(5), 697–710 (2002)
13. Jing, X.Y., Yao, Y.F., Zhang, D., Yang, J.Y., Li, M.: Face and palm print pixel level fusion and Kernel DCV-RBF classifier for small sample biometric recognition. Pattern Recognition 40, 3209–3224 (2007)

14. Hwang, H.: Daily Electric Load Forecasting Based on RBF Neural Network Models. International Journal of Fuzzy Logic and Intelligent Systems 13(1), 37–46 (2013)
15. Ha, S.-H., Jeon, H.-T.: Development of Intelligent Gear-shifting Map Based on Radial Basis Function Neural Networks. International Journal of Fuzzy Logic and Intelligent Systems 13(2), 116–123 (2013)
16. Na, J.H., Park, M.S., Choi, J.Y.: Linear boundary discriminant analysis. Pattern Recognition 43, 929–936 (2010)
17. Cervantes, A., Galvan, I.M., Isasi, P.: AMSPO: A New Particle Swarm Method for Nearest Neighborhood Classification. IEEE Transaction on Systems, Man, and Cybernetics Part B 39(5), 1082–1091 (2009)

Recursive Neuro-Fuzzy Algorithm for Flow Prediction and Pump On-Off Minimization

Ho-Hyun Lee[1], Sang-Bok Jang[1], Gang-Wook Shin[1], Sung-Taek Hong[1], and Myung-Geun Chun[2,*]

[1] K-water Institute, 125,1689Bean-gil, Yuseong-daero, Yeseong-gu, Daejeon, 305-730, Korea
dlgugus@gmail.com, {jsbok0502,gwshin,sthong}@kwater.or.kr
[2] Control and Robot Engineering, Chungbuk National University,
52 Naesudong-ro, Heungdeok-gu, Cheongju Chungbuk 361-763, Korea
mgchun@chungbuk.ac.kr

Abstract. In the water treatment process, a main objective is to improve the water quality and also minimize the production costs. To achieve these, an integrated monitoring and control system has been established through flow prediction and pump scheduling. This paper proposes a new integrated solution for predictions and optimal pump control by learning algorithms. Flow prediction has usually been studied for daily or monthly estimation, which is insufficient for real-time control of a water treatment plant (hereafter WTP). An hourly based estimator is proposed to track the steady change of flow demand. Unlike electricity, water can be stored in huge tanks for more than a dozen hours, which can be used for saving energy and increasing water quality. Pump on/off minimization is considered to improve the water quality. If influent water to a water treatment plant varies, then output turbidity and particles are increasing, which could possibly be supplied to citizens. The proposed on/off minimization algorithm is expected to prevent those particles from leaking and to secure public health.

Keywords: Water Treatment Process, Recursive Learning Algorithm, On/Off Minimization.

1 Introduction

Korea Water Resources Corporation is a public corporation that provides wholesale water to local water suppliers across Korea. K-water is responsible for the day-to-day transport of high-quality water from large headwater storage reservoirs to a series of junctions where local companies distribute the water to consumers.

In order to supply water efficiently from the headwater reservoirs to the retailers, estimates of consumer demands are generally required 48 hour in advance. This paper describes a computer-based mathematical model which relies on past demand data and weather forecasts to estimate consumption 48 hour in advance [1].

Most research papers have focused on the forecasting of urban water demand by considering annual, monthly or daily data [2][3]. Such long-periods of flow

* Corresponding author.

Y.I. Cho et al. (eds.), *Soft Computing in Artificial Intelligence*,
Advances in Intelligent Systems and Computing 270,
DOI: 10.1007/978-3-319-05515-2_11, © Springer International Publishing Switzerland 2014

predictions are usually easier to estimate than shorter periods like hourly flow predictions. Even though hourly predictions are much more difficult and are likely to have more errors, they need to be estimated in order to control the pumps and valves efficiently because they can reduce energy consumption and improve the water quality with well organized load balancing [4]. By this method, prediction errors will be minimalized and WTP operators will be able to adjust the flow more precisely based on trends in citizen's flow demand.

Usually pumps have been operated based on an operator's experience, necessities and convenience. Instead of relying on an operator's heuristic method, the usage of pumps should be determined using statistics such as estimated flow and level, which can minimize the trials of pump on and off that influence the relevant facilities and the produced water quality. Fig. 1 shows that water quality usually deteriorates when pump operations change. Thus, water treatment plants must reduce the number of pump operation changes as much as possible.

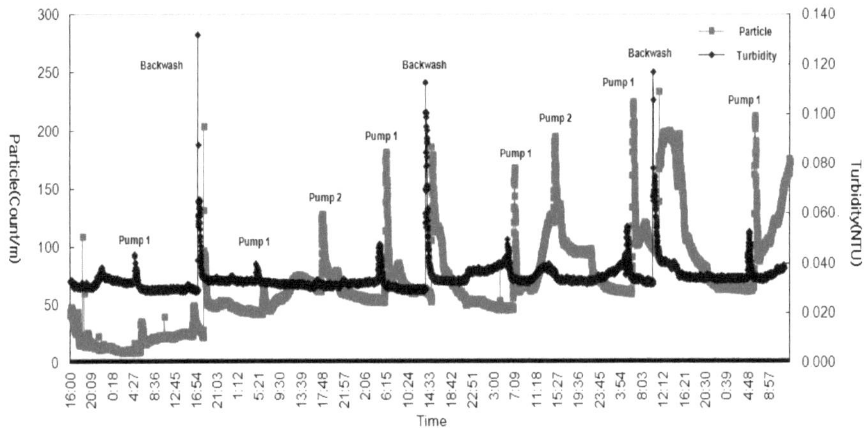

Fig. 1. Water Quality fluctuations as a result of Pump Operation Changes

2 Simulation and Experiment

A case study has been carried out at a WTP in Korea. The flow rate of this treatment plant is about 90,000 cubic meters per day. The plant consists of one influent pump station, one water treatment plant and dozens of storage tanks. The output flows of tanks are scattered here and there. Flow prediction is estimated by five tank output flows that are equal to 77 percent of the total tanks. Three of them have output flow meters while the others don't have, in which output flow are indirectly calculated by input flow and level.

2.1 Simulation

2.1.1 Learning Algorithm

Widely used prediction algorithms such as regression, neural network and support vector machine are considered and their results are shown in Table 1. As you can see, the linear regression has a large error that is caused by the non linear characteristics of the target process, or hourly flow prediction [5]. Instead of linear regression, a polynomial kernel [6] could be applied to learn non-linearity. Compared to the linear regression, its result is quite reasonable. The multi layer perceptron shows better results as hidden layers are increased, in which the neural networks are more complicated to implement [7][8]. ANFIS is also a kind of neural network which usually uses a back propagation algorithm [9]. Its results are almost similar to MLP and fuzzy C, meaning that clustering is also used to reduce the learning parameters [10][11]. LS-SVM takes a long time to calculate, in which all of the training data are considered as support vectors. Even though its results are the best, it is not recommended to have too many parameters to learn [12].

Table 1. Comparison of Learning Algorithms

Error		Linear Regression	MLP (Hidden 20)	ANFIS	Support Vector Regression
Training Error	MAPE	22.75	12.49	12.80	12.22
	MSE	178,050	68,841	71,856	67,186
Test Error	MAPE	22.67	12.87	12.95	12.54
	MSE	179,070	72,572	74,107	71,052
Num. of Parameter		4	4*20+21*1=101	14*4=56	18,950

As hourly flow data is affected by prior information, recursive function to call itself, is considered and previous flows k(t-1), k(t-2) and k(t-3) are added for auto regression. Table 2 shows that the learning algorithm with recursive form achieved a better result and it is combined with MLP and neuro- fuzzy algorithm. Even though MLP is better than neuro-fuzzy, neuro fuzzy is easier to implement because it has less parameters. Therefore, the recursive neuro-fuzzy algorithm is chosen to be a learning algorithm.

Table 2. Comparison of Learning Algorithm with Recursive Form

Error		AR	LR with AR	MLP with AR (Hidden 20)	ANFIS with AR
Training Error	MAPE	17.23	16.27	11.53	12.57
	MSE	115,160	107,500	60,693	70,302
Test Error	MAPE	17.18	16.45	11.75	12.73
	MSE	117,140	110,630	62,791	73,696

Results of Flow Prediction

Roughly 2,300 sample data from June to August were collected from SCADA(Supervisory Control and Data Acquisition) system to check the accuracy of flow prediction. Fig. 2 shows the learning results by recursive neuro-fuzzy algorithm. The flow rate is less than 1,300m3/h early in the morning and maximum flow is as high as 3,600m3/h. Around 20 hundred hours (8pm), the flow rate has another peak. The algorithm follows the real flow well, thus, it expected to be reliable for control pump minimization.

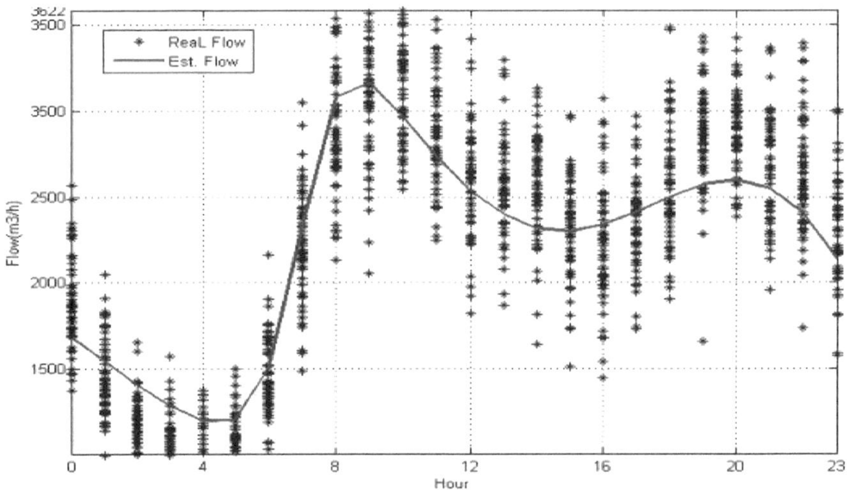

Fig. 2. The Results of Hourly Flow Prediction

2.1.2 On-Off Minimization

Sample data are used from June to August last year, when data loss was minimal. Fig. 3 describes the flow curve according to when the pump was on or off. Pump repeats on/off every 5 hours on average.

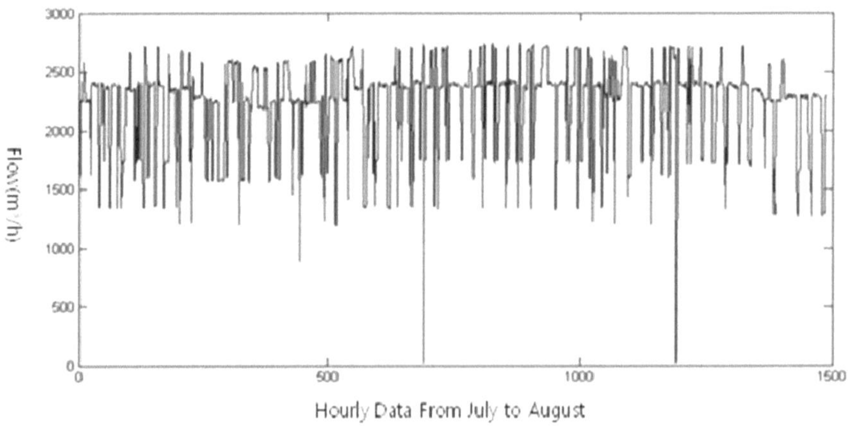

Fig. 3. The On-Off Status of Pumps

The On/Off minimization algorithm is closely related with equal distribution to controllable tanks, through which storage tanks can be used maximally between min and max levels. It leads to pump on/off every 74 hours on average which is equal to 20 times over a period of 1488 hours. This is shown in Fig. 4.

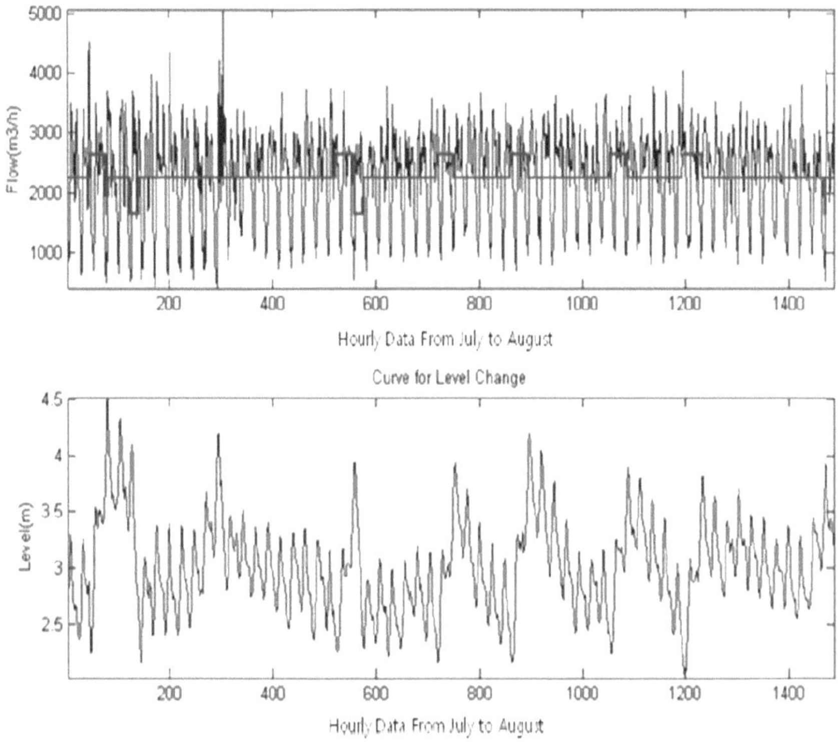

Fig. 4. The Number of Pump On/Off & Level Estimation for the Proposed System

2.2 Experiment

2.2.1 Flow Prediction

This experiment was taken for 24 hours. At that time, the considered variables were Tuesday and Wednesday and the temperature was 5 degree. Fig. 5 shows the curve for pump, real and predicted flow according to the time. Its estimation result has 7.3 percentage of mean absolute percentage error, which might come from On-Off control of a big customer such as a company or an apartment.

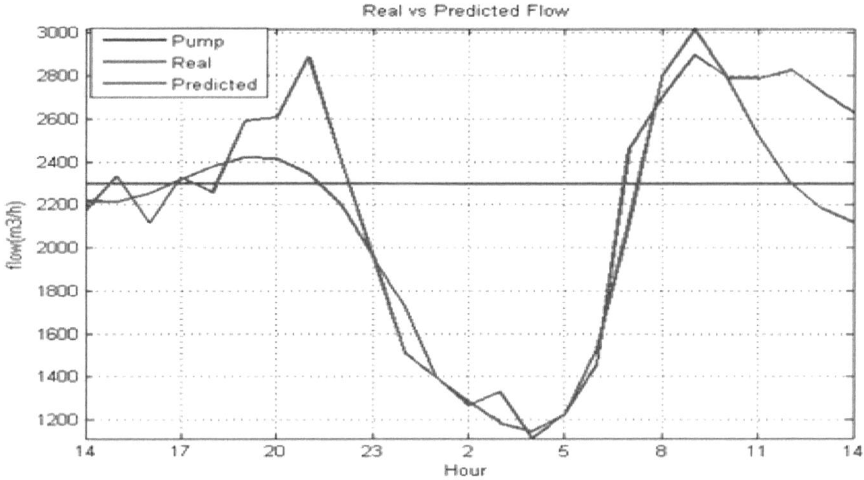

Fig. 5. Experimental Result of Real vs Estimated Flow

2.2.2 Pump On-OFF Minimization

It was supposed that 40 percent of the total level, or 2 ~ 4m, could be used, the varying curve from the minimum to maximum levels are shown in Fig. 6. The maximum daily level difference is located between 22 hundred hours (10pm) and 7 hundred hours (7am), which is almost 1m in a day. As the daily level difference in the above figure is just 60 cm in the case that flow demand is roughly 50,000m^3/d, the pump would reach its maximum operation level at 5 hundred hours (5am) on the third day, which means that the pumps can operate for 63 hours continuously without turning them off.

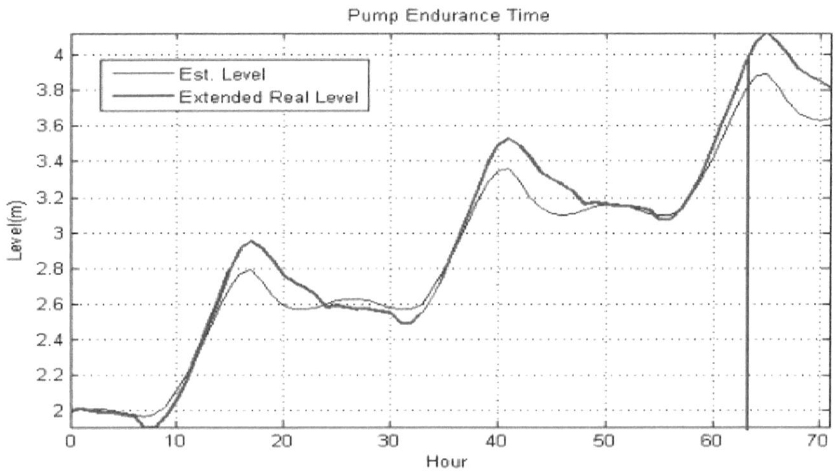

Fig. 6. Pump Operation Time without Change

3 Conclusion

In the water treatment process, the present target is to provide an integrated solution to predict flow and control the pumps and valves for water quality and to minimize costs. To achieve these, an integrated monitoring and control system has been established through flow prediction and pump scheduling. Because of the lack of models and the controllers of plants, it has the limitation to reach the target in Korea. Thus, learning algorithms are proposed to approximate unknown plants without expert knowledge.

First, flow predictions are estimated by learning algorithm with recursive form, which makes it possible to track the latest trends in demand and it is not necessary to take any additional learning. The Proposed algorithm, recursive neuro-fuzzy, has less parameter to be learned and better accuracy than other algorithms. It results in the easy implementation to water treatment plant.

Second, pump on/off minimization is proposed to increase water quality and reduce damages to electrical and mechanical facilities. If influent water to a water treatment plant varies, then output turbidity and particles are increasing, which could possibly be supplied to citizens. The proposed on/off minimization algorithm is expected to prevent those particles from leaking and to secure public health by supplying hygienically safe drinking water.

References

1. Parker, J.M., Wilby, R.L.: Quantifying Household Water Demand: A Review of Theory and Practice in the UK. Water Resources Management 27(4), 981–1011 (2013)
2. Msiza, I.S., Nelwamondo, F.V., Marwala, T.: Water Demand Prediction using Artificial Neural Networks and Support Vector Regression. Journal of Computer 3(11) (November 2008)
3. Zhou, S.L., McMahon, T.A., Walton, A., Lewis, J.: Forecasting daily Urban water demand: a case study of Melbourne. Elsevier Journal of Hydrology 236, 153–164 (2000)
4. David, V.B., Joseph, A.Y., Eileen, M.W., Simon, B.: Implementing an Energy Management System for Distribution Pumping. In: AWWA Anual Conference and Exposition in San Francisco, CA (2005)
5. Gauss, C.F.: Theoria Motus Corporum Coelestium in Sectionibus Conicis Solem Ambientum (1809)
6. Vapnik, V., Golowich, S., Smola, A.: Support vector method for multivariate density estimation. In: Advances in Neural Information Processing Systems, vol. 12, pp. 659–665. MIT Press, Cambridge (1999)
7. Alpaydin, E.: Introduction to machine learning, pp. 229–251. MIT Press (2004)
8. Rumelhart, D.E., Hinton, G.E., Williams, R.J.: Learning internal representations by error propagation. Cognitive Science, 4–8 (1985)
9. Jang, J.-S.R.: ANFIS: Adaptiv-Network-Based Fuzzy Inference System. IEEE Transactions on Sytem, Man and Cybernetics 23(3), 665–685 (1993)

10. Bezdek, J.C., Ehrlich, R., Full, W.: The Fuzzy c-Means clustering. Elsevier Computers and Geosciences 10(2-3), 191–203 (1984)
11. Talei, A., Hock, L., Chuab, C., Quekc, C., Janssond, P.-E.: Runoff forecasting using a Takagi–Sugeno neuro-fuzzy model with online learning, vol. 488, pp. 17–32. Elsevier (2013)
12. Valyon, J., Horvath, G.: A Robust LS-SVM Regression. World Academy of Science, Engineering and Technology 7 (2005)

Logotype Design Support System Based on Sketches

Junji Saikawa[1] and Takehisa Onisawa[2]

[1] Graduate School of Systems Information and Engineering, University of Tsukuba
1-1-1, Tennodai, Tsukuba 305-8573, Japan
tkb20828@fhuman.esys.tsukuba.ac.jp

[2] Faculty of Engineering, Information and Systems, University of Tsukuba
1-1-1, Tennodai, Tsukuba 305-8573, Japan
onisawa@iit.tsukuba.ac.jp

Abstract. The paper proposes a support system for interactive logotypes design based on user's sketches, which applies Interactive Genetic Algorithm. A user inputs a letters string and a rough sketch of a layout of a desired logotype. According to user's inputs, the system presents design candidates to a user. Then, a user evaluates them. A user can also modify the logotype design candidates by rough and simple sketches as necessary. According to the evaluation and the modification, the system has other new design candidates. The procedures are repeated until a user is satisfied with presented design candidates. This paper also explains two types of experiments for the confirmation of proposed system validity: whether the system supports logotype design, and whether sketch inputs are useful. Experimental results show the usefulness of the proposed system.

Keywords: logotype design, sketch, interactive genetic algorithm, shape transformation.

1 Introduction

Recently, many support systems or generation systems are proposed using the evolutionary computation method, which generate creative works such as logotype design [1], sound generation [2], layout design [3], shape design of a mobile phone [4], color planning [5], repeating the interaction between a user and a computer agent until a user is satisfied with these works. These systems are called an evolutionary design system. A user can have creative design by evaluation of presented design candidates in the interaction process. However, the evaluation is often performed using a numerical evaluation value based on his/her preference. This means that a user takes part in design not actively but passively. Then, other evolutionary design methods by which a user takes part in design actively is desired. Furthermore, some of above systems deal with user's subjective Kansei information, and user's impression of desired works is expressed using impression words [6-9]. Impression of non-verbal media such as music, image, however, is not necessarily expressed linguistically. If it is difficult to express the impression linguistically, the systems

Y.I. Cho et al. (eds.), *Soft Computing in Artificial Intelligence*,
Advances in Intelligent Systems and Computing 270,
DOI: 10.1007/978-3-319-05515-2_12,

cannot have appropriate information of the impression of desired creative works and cannot generate creative works reflecting user's impression of the works. Other ways of expression of the impression should be also considered.

There are researches using sketches such as studies on image retrieval using sketches as queries [10-11], 2D or 3D sketch graphic design systems [12-14], car body design [15]. In [15], the design system presents sketches as car body design candidates. According to [16], the process of sketching can support the sort of transformational thinking that is seen as essential for the interpretation and reinterpretation of ideas in innovative design. Input by sketches in design systems is more useful for creative design works than that by words.

This paper proposes an interactive logotype design support system that uses user's drawing sketches as information on user's impression of desired logotype design. That is, the sketching process is introduced to the evolutionary design procedures as the new type of evolutionary interactive design system, and a user can take part in design actively. A sketch is usually a drawing of a view, a still life, etc. In this paper, however, a sketch is a trial drawing of a logotype layout with a single color and simple lines. According to [17], a logotype designer draws many rough sketches or makes design candidates base using existing fonts, and fixes his/her impression by modifying the candidates base. These tasks, however, are very difficult for a design beginner who does not know how to design logotypes reflecting his/her impression of desired logotypes. Then, this paper proposes the following logotype design framework. A user draws rough sketches of layouts of desired logotypes. The support system designs logotypes candidates based on sketches and presents them to a user. A user evaluates and modifies the candidates using numerical values and newly drawing sketches. The evaluation/modification procedures are repeated until a user is satisfied with logotype design. The proposed system uses so simple and rough sketches of logotype that even a beginner of logotype design can draw.

A font generation system applying interactive genetic algorithm is proposed [18]. This system consists of the font definition phase and the font creation phase, and experimental results show good performance. However, this study [18] considers only the font face and the font size as a font design. On the other hand, a logotype design is different from a font design from the following viewpoints: An image to be impressed to other people is expressed by a logotype. Therefore, in logotype design, not only design of each letter face and the size of each letter but also the shape of each letter and the layout of each letter are needed to be considered. Therefore, this paper does not use existing font types.

The organization of the paper is as follows. Section 2 explains designed logotypes in this paper. Section 3 shows the structure of the proposed support system and explains the design procedures. Section 4 describes subject experiments performed in order to show the usefulness of the design support system and shows their results. Final section shows conclusions of this paper.

2 Logotypes and Transformation Operations in Design

A logo is usually a logotype, a symbolic mark or a logo mark, and a company name, an organization name, or a trade name is expressed by their combination. In this paper, a logotype is defined as designed alphabetical letters and in logotype design not only design of each letter face and the size of each letter but also the shape of each letter and the layout of each letter are considered. Therefore, many types of transformations of letters are considered. As for the color scheme of a logotype, background color and letters color are assumed to be white and black, respectively, for design simplification.

2.1 Letters

Alphabetical letters composing logotypes are composed of straight lines (a horizontal line, a vertical line, an oblique line) or a curve line, and each line has five points as shown in Fig.1(1), where these points have coordinate values in an x-y plane. This type of a letter is called a skeleton-typed letter in this paper. A skeleton-typed letter is made thick as shown in Fig.1(2). This type of a letter is called a bold-typed letter in this paper.

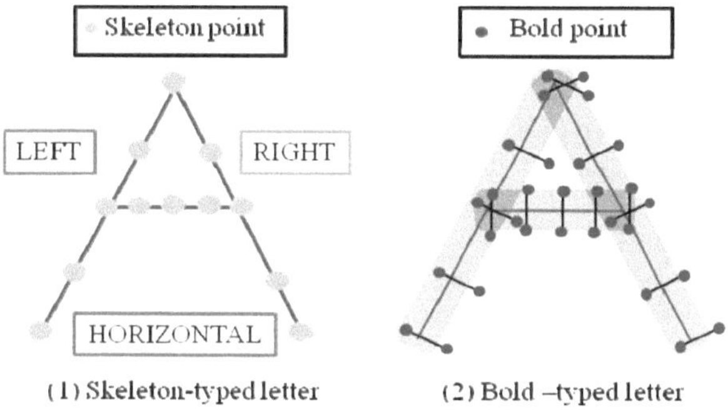

Fig. 1. Letters

2.2 Letter Transformation Operations

Each letter has letter parameters, and lines composing a letter have line parameters as shown in Table 1. Logotypes are designed by transforming lines composing a skeleton-typed letter. Transformation procedures are as follows. (1) The basic composition of a letter is made and a skeleton-typed letter is generated. (2) Each line of a letter is transformed and a skeleton-typed letter is made thick. Then, a bold-typed letter is generated. (3) The center of gravity of a letter is obtained, and expansion, reduction and/or obliqueness of a letter are performed.

Basic Composition of a Letter. The basic composition of a letter is made by the aspect ratio and the parameter values of x-axis and y-axis base lines of a letter, where the aspect ratio is fixed within the range of the ratio of the length and the width that is from 100[pix]x50[pix] (2:1) to 50[pix]x100[pix] (1:2) compared with a square with 100[pix]x100[pix]. Fig.2 shows examples of letters ‘O’ and ‘R’. The coordinate values of each line of ‘O’ or ‘R’ are obtained by the coordinate values of the points of a rectangular (P_1, P_2, P_3, P_4) that are determined by the aspect ratio, and the parameter values of base lines $(xBase, yBase)$.

Table 1. Design parameters

Parameter Types	Parameter Names		Explanation
Letter Parameters	Aspect xBase, yBase xShift, yShift centroid scaling		aspect ratio of letter x-axis and y-axis base line of letter slant ratio of letter center of gravity of letter scaling ratio of letter
Line Parameters	Skeleton-typed letter parameters	changeFlag hv dia_a, dia_b cur_a, cur_b	shape change flag shape change ratio of horizontal or vertical line shape change ratio of oblique line shape change ratio of curver line
	Bold-typed letter parameters	interpolationStyle edgeStyle widthStyle startWidth, endWidth	inerpolation style edge style width style start or end width

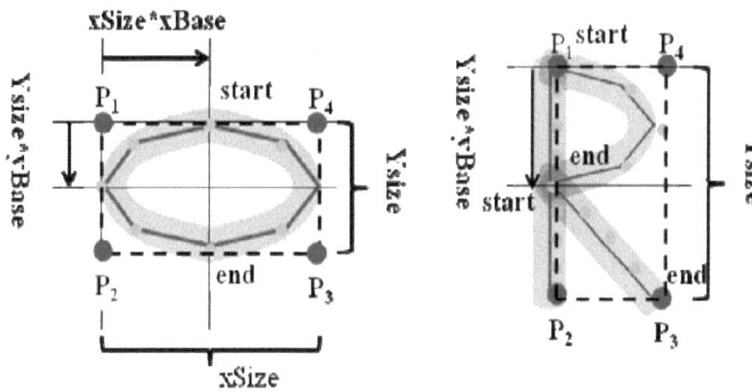

(1) Aspect ratio=-50 (2) Aspect ratio=50

Fig. 2. Examples of basic composition of letter

Transformation of a Line. The transformation of a line is the transformation of each line composing a letter. In this operation the movement operation of the points on each line of a skeleton-typed letter and the generation of a bold-typed letter from a skeleton-typed letter are performed.

Movement of Points on Line of Skeleton-typed Letter. This operation determines the coordinate values of points on a line of a skeleton-typed letter. The middle point Q_3 of five points $(Q_1, Q_2, Q_3, Q_4, Q_5)$ on a line of a skeleton-typed letter is determined by the parameter values of base lines $(xBase, yBase)$ as shown in Fig.3(1). Furthermore, the coordinate values of point Q_2 between Q_1 and Q_3, and those of point Q_4 between Q_3 and Q_5 are obtained according to the type of a line as follows. If a line is a vertical line or a horizontal line, the movements of Q_2 and

Q_4 are controlled by parameter *hv* as shown in Fig.3(2). If a line is an oblique line, the movements of Q_2 and Q_4 in X axis direction are controlled by parameter *dia_a* and those in Y axis direction are controlled by parameter *dia_b* as shown in Fig.3(3) and Fig.3(4). If a line is a curve, the movements of Q_2 and Q_4 in X axis direction are controlled by parameter *cur_a* and those in Y axis direction are controlled by parameter *cur_b* as shown in Fig.3(5).

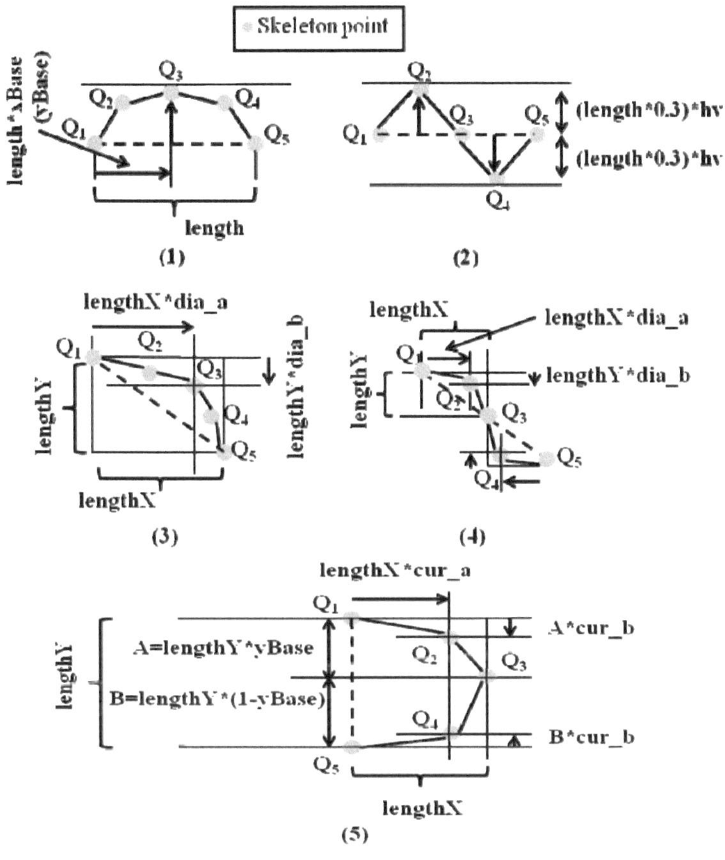

Fig. 3. Example of points movement

Thickness of Line of Bold-typed Letter. The thickness of one endpoint of a line of a bold-typed letter and that of another endpoint of a line of a bold-typed letter are determined by parameters *startWidth* and *endWidth* as shown in Fig.4. The change of thickness between two endpoints is determined by parameter *widthStyle*.

Interpolation of Coordinate Values of Bold-typed Letter. The style of interpolation between two points in a bold-typed letter, i.e., interpolation by a straight line or that by a curve, is determined by parameter *interpolationStyle*.

Shape of Edge of Bold-typed Letter. The shape of edge of a line of a bold-typed letter, i.e., a straight or a curve, is determined by parameter *edgeStyle*.

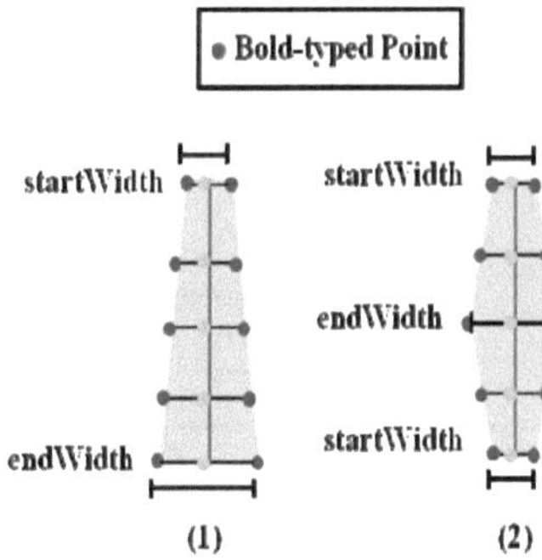

Fig. 4. Thickness of line

Transformation of Letter. The transformation of a letter is two dimensional projective transformation of the coordinate values of 4 points of a rectangular surrounding a skeleton-typed letter that are determined by the aspect ratio. The projective transformation is defined by Eqs. (1) and (2).

$$x' = \frac{a_1 x + b_1 y + c_1}{a_0 x + b_0 y + 1}, \tag{1}$$

$$y' = \frac{a_2 x + b_2 y + c_2}{a_0 x + b_0 y + 1}, \tag{2}$$

where (x, y) are the coordinate values of one of 4 points surrounding a skelton-typed letter that are determined by the aspect ratio, (x', y') are the coordinate values of the corresponding 4 points surounding a skeleton-typed letter after two dimentional projective transformation, and $(a_0, a_1, a_2, b_0, b_1, b_2, c_1, c_2)$ are coefficients that are obtained by data of pre-transformed letters and data of transformed letters. The following three types of letter transformations are considered.

Setting Center of Gravity of Letter. This operation fixes the coordinate values of the center of gravity of a rectangular with four points (P_1', P_2', P_3', P_4') surrounding a transformed letter by moving the coordinate values of the center of gravity of a rectangular with four points (P_1, P_2, P_3, P_4) surrounding a pre-transformed letter as shown in Fig.5(1).

Expansion or Reduction. This operation expands or reduces a letter. This operation fixes the coordinate values of four points (P_1', P_2', P_3', P_4') of a rectangular surrounding an expanded or reduced letter by multiplying *xSize* and *ySize* of a rectangular surrounding a pre-expanded or pre-reduced letter by *scaling* as shown in Fig.5(2).

Obliquesness of Letter. This operation slants a letter in the x-axis or the y-axis direction. This operation fixes the coordinate values of four points (P_1', P_2', P_3', P_4') of a rectangular surrounding a slanted letter by multiplying *xSize* in the x-axis direction by *xShift* or by multiplying *ySize* in the y-axis direction by *yShift* of a rectangular surrounding a pre-slanted letter as shown in Fig.5(3), where this operation is limited to the slant of a letter in only one direction.

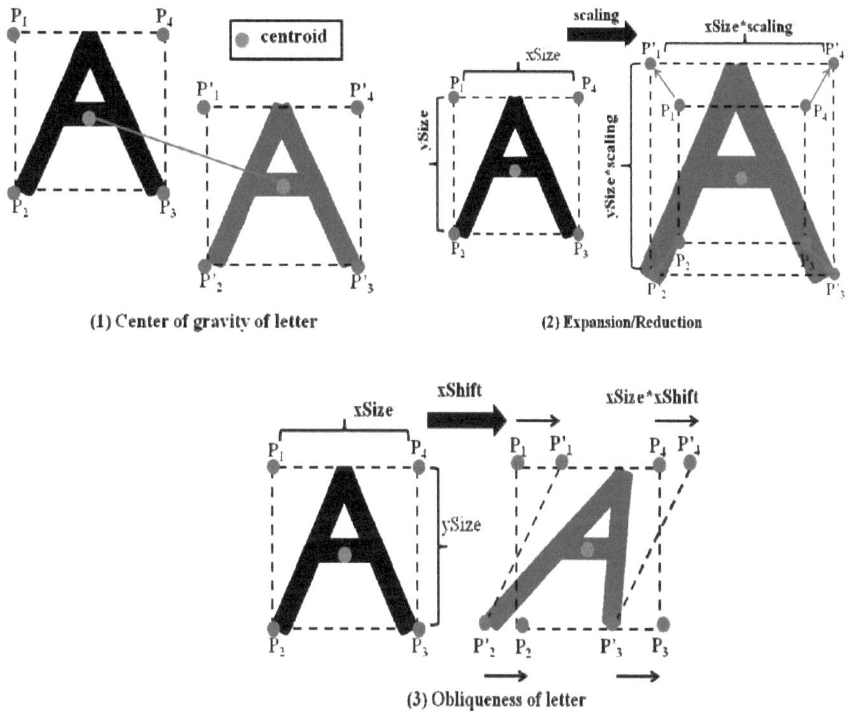

Fig. 5. Transformation of letter

3 Design Support System

The proposed system generates logotypes design candidates based on user's sketches of desired logotypes by the interaction with a user. The system applies Interactive Genetic Algorithm (IGA) [19]. In the interaction a user can modify the shape of a letter and/or the shape of a line of a letter by rough sketches as necessary.

Before starting design, a user has impression of a desired logotype design. Then, a user inputs a letters string of desired logotypes and a rough sketch of its layout. The system generates design candidates based on the sketch, and presents them to a user. A user evaluates them subjectively whether they reflect a user's impression of a desired logotype design. Besides subjective evaluation, a user also modifies design candidates by rough and simple sketches of a letter shape and/or a line shape of a letter. The system generates new design candidates according to user's evaluation and user's modification by sketch drawing, and presents new design candidates to a user again. The procedures of system designing and presentation of design candidates, and user's evaluation and drawing sketches are repeated until a user is satisfied with designed logotypes. Fig.6 shows the flow of the system.

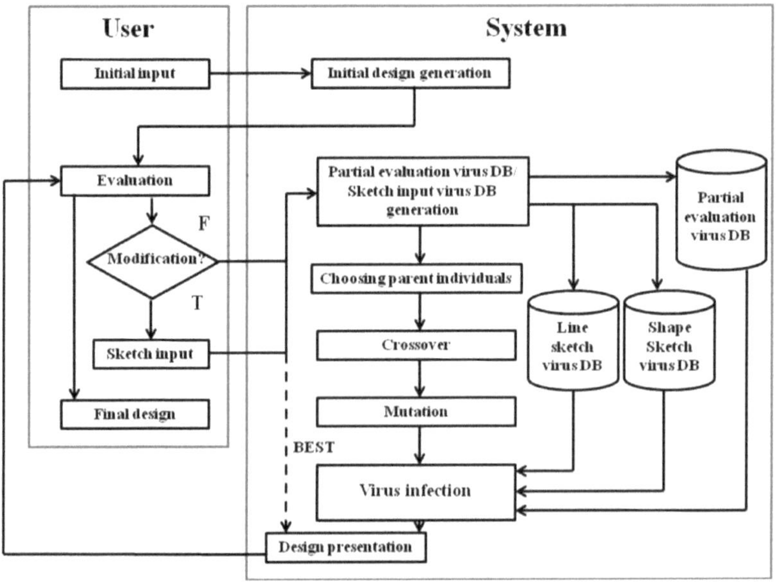

Fig. 6. Flow of system

3.1 Chromosomes

The parameters mentioned in Section 2 are represented by chromosomes in this system. Fig.7 shows an example of chromosomes taking a letters string 'TEST' as an example. A chromosome consists of letter parameters and line parameters, and each line parameter has a numerical value according to the type of a line as arrows shown in Fig.7. The position of the center of gravity of each letter *centroid* and expansion/ reduction rate *scaling* have fixed parameter values that are inputted as initial values, and these values are not changed in GA operations. The uniform crossover operation is used as the crossover operation. The individuals chosen at random get the mutation operation by probability $p_{mutation}$. The proposed system also uses the elitist preserve strategy.

3.2 Generation of Initial Individuals

Except for *centroid* and *scaling*, all parameters are fixed at random as initial values. A user inputs a letters string of desired logotypes and draws a rough and simple sketch of a layout of desired logotypes on the canvas window using a mouse as shown in Fig.8, where this canvas window is designed as an input interface, and one stroke in the sketch of the layout is dealt with as one letter. A rectangular surrounding a drawing area of one stroke sketch is extracted. Coordinate values of the center of gravity of the extracted rectangular are obtained as *centroid*. Expansion/reduction rate *scaling* is obtained by the comparison of the length of a longer line of the rectangular and the basic size 100[pix].

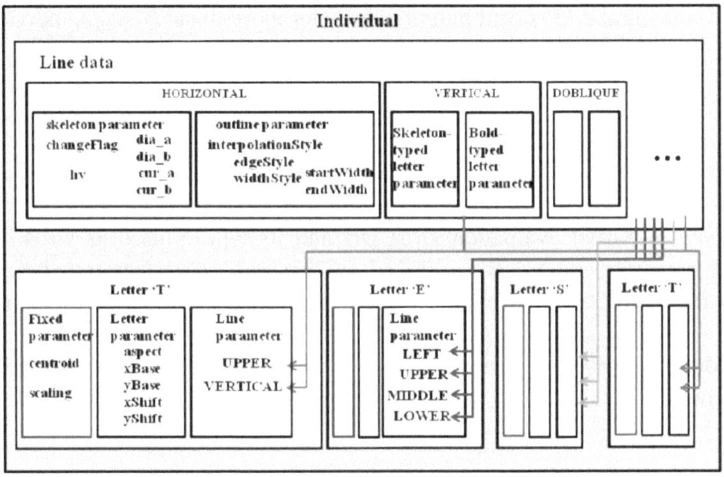

Fig. 7. Example of chromosomes

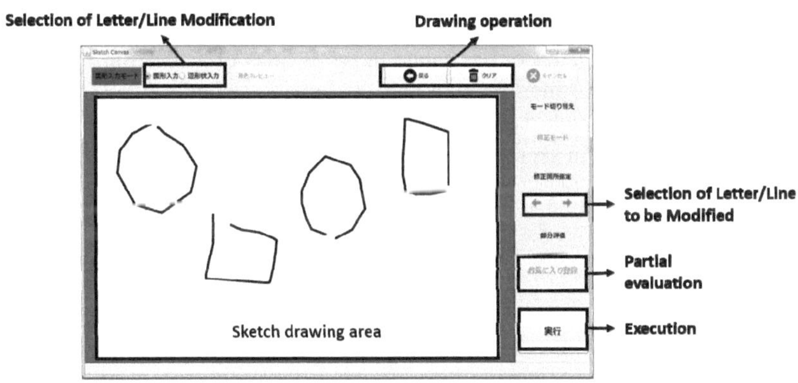

Fig. 8. Rough sketch of layout

3.3 User's Evaluation

A user evaluates presented design candidates subjectively. Whole evaluation and partial evaluation are considered as user's evaluation. The whole evaluation has two types of evaluations for a design candidate. In one evaluation a user gives the evaluation value to a presented design candidate with a 4-levels scale such as -1 (bad), 0(neutral), +1(good), +2(very good). In the other evaluation a user gives the best evaluation among presented design candidates at each generation. In this system the fitness value is defined by user's evaluation value as follows; the best:10, very good(+2):5, good(+1):3, neutral(0):1, bad(-1):0. In the GA operations presented individuals are copied up to their fitness values and are saved in the next generation parent candidates pool, where the total number of saved individuals is limited to N_{parent}. If the number of saved individuals is not more than N_{parent}, individuals are chosen from the current generation parent candidates pool at random and saved in the next generation parent candidates pool up to N_{parent}, where the fitness value of newly saved individual is defined as 1.

The partial evaluation is the evaluation of part of a bold-typed letter. If a user likes some type of a transformation of a line among four types mentioned in 2.2, a user can save its type in partial evaluation virus DB and its type is used as virus infection, where this reservation in DB is limited to one per one generation because this operation has a direct influence on the design generation. Virus infection is mentioned in 3.5.

The individual with the best evaluation is preserved as an elite individual at the next generation.

3.4 Modification by Sketch

A user can modify presented design candidates by rough and simple sketches drawn on the canvas window directly using a mouse as shown in Fig.9, which is the same window as the one shown in Fig.8. The modification by sketch is to change parameter values of a skeleton-typed letter or to change those of lines of a skeleton-typed letter. The former is called a letter modification and the latter is called a line modification. At the modification by sketch, however, only one of two modifications is possible. This modification is saved in sketch virus DB and is used at the virus infection.

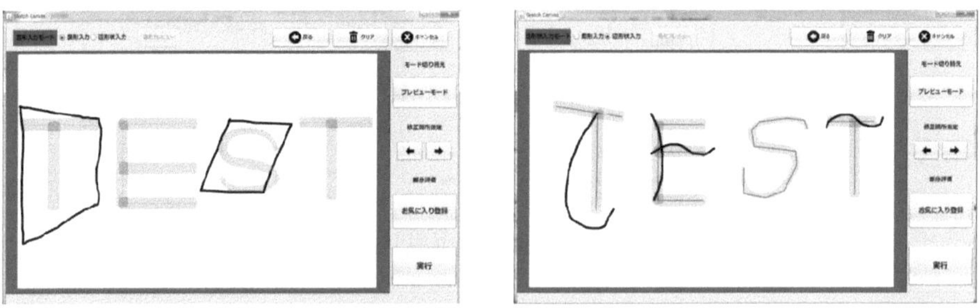

(1) Rough sketch for letter modification (2) Rough sketch for line modification

Fig. 9. Modification by rough sketch

Letter Modification. This operation is to modify the shape of a chosen letter by drawing the rectangular surrounding the chosen letter. This operation is limited to the design candidate with the best evaluation because this operation has a direct influence on the generation of design candidates. A user chooses a letter to be modified and draws a rectangular surrounding the chosen letter. The coordinate values of the four points (P_1, P_2, P_3, P_4) of the drawn rectangular in this operation are saved in letter modification sketch virus DB. Fig.10 shows the flow of this operation.

Line Modification. This operation is to draw a line on a chosen letter directly. A user chooses a line of a letter to be modified and draws the shape of a line on the letter. The number of lines to be modified has no limitation. The rectangular surrounding the drawn line is obtained, and the expansion or reduction rate is obtained by the drawn rectangular. The drawn line and the expansion or reduction rate are saved in line modification sketch virus DB. Fig.11 shows the flow of this operation. If a letter is modified in both letter modification and line modification such as 'T' shown in Figs.10 and 11, this letter is modified by the line modification after the letter modification. And the calculation of scaling is not performed because in this case the operation of the movement of points on the line is not performed in virus infection.

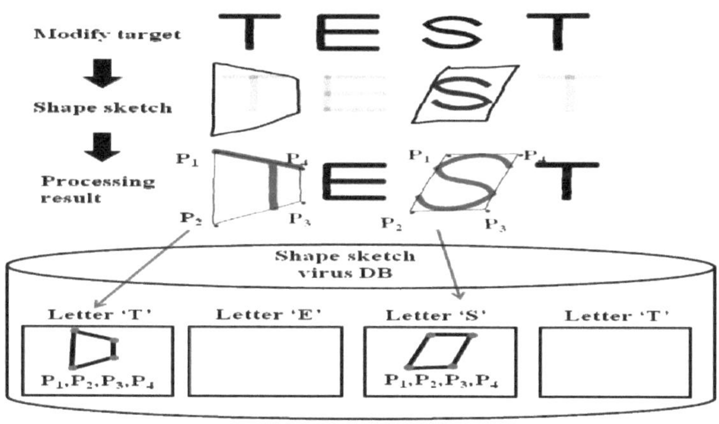

Fig. 10. Letter modification and its sketch virus DB

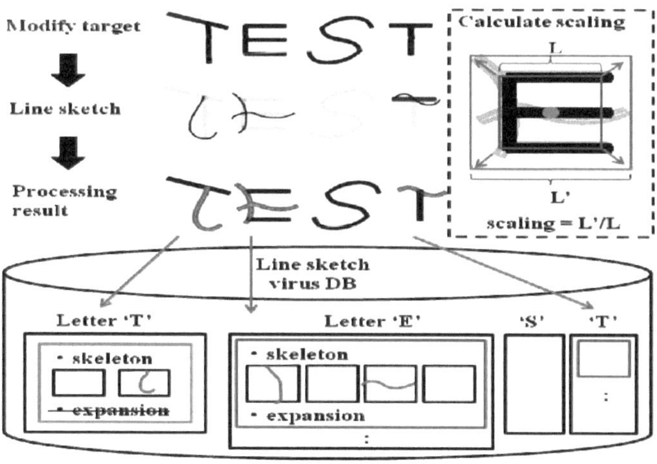

Fig. 11. Line modification and its sketch virus DB

3.5 Virus Infection

The operation of virus infection is performed using viruses saved in each virus DB in order to reflect user's partial evaluation and the modification by sketch. The partial evaluation and the modification by sketches of logotypes are reflected in the GA operators by virus generation and virus infection. Part of a bold-typed letter, to which a user gives good evaluation, and the modified letter or the modified line, which a user modifies by sketches, play a virus role. After the crossover and the mutation, part of a logotype is infected with viruses, i.e., viruses are copied to the part of a logotype. The infected part of a logotype is the same as that of the virus. The usefulness of virus generation and virus infection is confirmed by [20].

Virus Infection by Partial Evaluation. According to the rate of virus infection by partial evaluation $p_{partialeval}$, the virus chosen from partial evaluation virus DB at random infects the following three groups. The one is the group of lines in skeleton-typed letters with user's partial evaluation. This group is infected with the virus with probability p_{target}. The second one is the group of skeleton-typed letters of which lines have user's partial evaluation. This group is infected with the virus with probability p_{group}. The other is the whole of a letter that is infected with the virus with probability p_{whole}. A virus has infectious capacity $N_{infection}$. This capacity decreases by 1 at every infection, and when the capacity is 0, the virus is deleted from DB.

Virus Infection by Letter Modification. If a letter is modified by the letter modification that is saved in letter modification sketch virus DB as a virus, the corresponding part of all presented design candidates is infected by the virus. That is, the modified line is drawn on the corresponding part without referring to the fixed center of gravity, the aspect ratio, the expansion/reduction rate and the parameter values of the slant of a letter.

Virus Infection by Line Modification. The virus chosen from line modification sketch virus DB at random infects the line of a letter without referring to the fixed parameter values.

4 Subject Experiments

Subject experiments are performed in order to verify the following two points: Whether the proposed system helps a beginner user design logotypes, and whether the sketch input by a beginner user oneself is useful for the reflection of user's impression of desired logotype design. Then, the following items are considered as questionnaire in the experiments. (1) Whether a user is satisfied with designed logotypes. (2) Whether designed logotypes reflect user's impression of desired logotypes design. (3) Whether generated design candidates are appropriate. (4) Whether the sketch input is useful for the reflection of user's impression of desired logotypes. In the experiments subjects perform two types of experiments: The one is the experiments in which subjects design logotypes using the proposed system, which is called *Experiment A*.

The other is the ones in which subjects design logotypes using the existing logotype design service [21] that does not use the sketch input. This type of experiment is called *Experiment B*. These two experiments are performed in random order. Furthermore, the same string of letters and the same impression of desired logotypes are assumed in both *Experiments A* and *B*.

Twelve males/females of twenties are subjects and their experiences of logotype design are a few times at most. Only younger subjects with little experiences of logotype design perform the experiments in order to confirm the followings: Whether a beginner user can design logotype using the proposed system or not. They design logotypes of *LOVE*, *WORLD*, or *DESIGN*. Same initial individuals are fixed in the experiments.

4.1 Outline of Experiments

Experiment A. The experiment procedures are as follows. (1) A subject has impression of desired logotypes design and inputs a rough and simple sketch of the layout of desired logotypes design. (2) The system presents design candidates to a subject and a subject evaluates them. And a subject modifies design candidates using newly drawing sketch as necessary. (3) The design procedures are repeated ten times at most until a subject is satisfied with design candidates. (4) A user answers the questionnaire shown in Table 2 with a 7-levels scale,-3:not at all, -2: not, -1: not a little, 0: neutral, +1: a little, +2: normal, +3:very much. (5) After a few days a subject answers design evaluation questionnaire shown in Table 3 about the best individual at each generation with a 7-levels scale, -3:not at all, -2: not, -1: not a little, 0: neutral, +1: a little, +2: normal, +3:very much.

Table 2. Questionnaire

No.	Questionnaire Items
1	Are you satisfied with designed logotype or not? 1:whole of design, 2:shape of letters, 3:layout of letters
2	Does designed logotype reflect your impression of desired design logotype or not? 1:whole of design, 2:shape of letters, 3:layout of letters
3	Can you have other better impression by presented design candidates or not? 1:whole of design, 2:shape of letters, 3:layout of letters
4	Do you feel that presented design candidates approach to your impression of desired design logotype or not? 1:whole of design, 2:shape of letters, 3:layout of letters
5	Do you feel that a rough sketch of desired design logotype is useful for the reflection of your impression to designed logotype or not? 1:whole of design, 2:shape of letters, 3:layout of letters

Table 3. Design evaluation questionnaire

No.	Questionnaire Items
6	Are you satisfied with presented design candidates or not?
7	Do presented design candidates reflect your impression of desired design logotype or not?

Experiment B. The experiment procedures are as follows. (1) A subject inputs a letters string to be designed. (2) A subject chooses the genre of logotypes such as *logoplate*, *pop*, *headline style*, prepared in the logotype design service [21]. (3) A subject chooses a font type. (4) A subject fixes the parameter values of a size, color, thickness of a font and the transformation operations, and puts on the design generation button. (5) The design procedures are repeated until a subject is satisfied with design candidates at most 10 times. (6) After the design procedures, a subject answers questionnaire shown in Table 2 with a 7-levels scale, -3:not at all, -2: not, -1: not a little, 0: neutral, +1: a little, +2: normal, +3:very much, where the item 5 is not considered in this experiment.

4.2 Experimental Results

Fig.12 shows examples of impression of desired design logotype, initial sketch and the final designed logotypes in *Experiment A* by some subjects, and the final designed logotype in *Experiment B* by the same subjects, where impressions in Fig.12 are shown only to compare the impressions with final designed logotypes. It is found that various types of logotype designs are obtained in *Experiment A*.

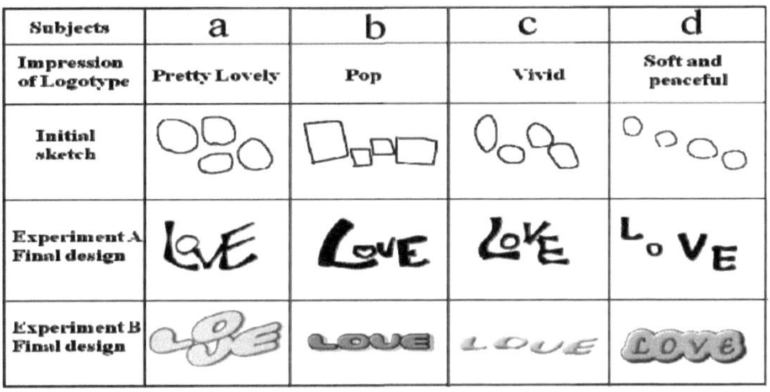

(1) LOVE

Fig. 12. Design examples in experiments

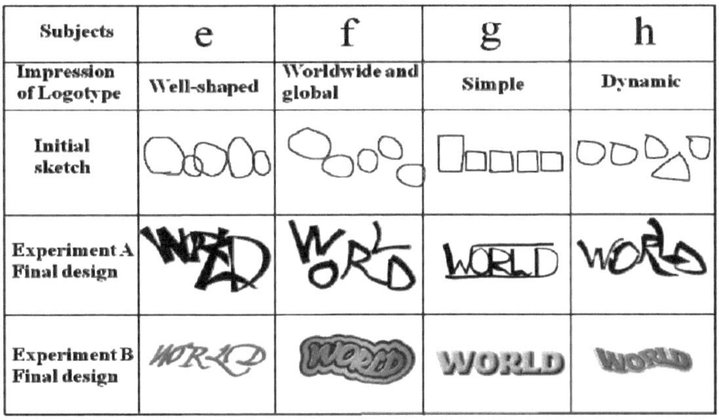

(2) World

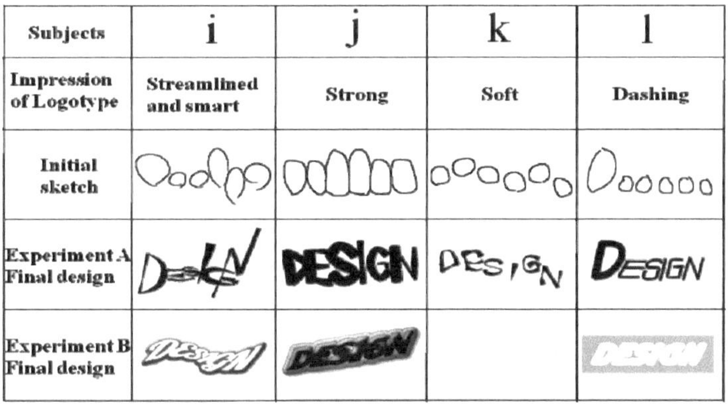

(3) DESIGN

Fig. 12. *(continued)*

Questionnaire Results in *Experiment A*. Fig.13 shows the average evaluation value among all subjects and 95% confidence interval estimation of the popular mean for subjects' evaluation for each questionnaire item. It is found from the average values and the confidence interval estimations that subjects give affirmative evaluation for each item. Table 4 shows the results of Wilcoxon rank sum test between the average evaluation values of the best design at the first generation and those of the best design at the 10th generation. It is found that subjects are satisfied with the designed logotypes because statistically significant difference is found between them.

Comparison Results of Questionnaire in *Experiment A* and *Experiment B*. Table 5 shows the questionnaire items that have statistically significant difference as the results of Wilcoxon rank sum test between the average value of each questionnaire item in Experiment A and that in Experiment B, where the items that are not shown in Table 5 have not statistically significant difference. It is found that subjects are more satisfied with the layout of logotype design in Experiment A than in Experiment B, and that subjects can have other better impression using the presented system than the existing logotype design service [21]. Then, the proposed system shows its usefulness.

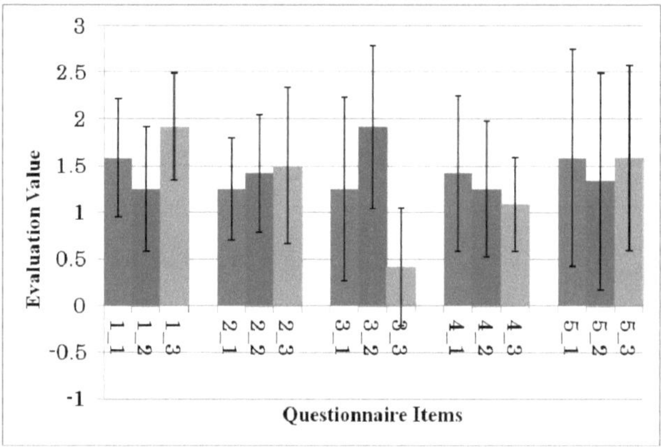

Fig. 13. Questionnaire results

Table 4. Comparison results of design evaluations

No.	Generation	Average Value	Significant Difference
6	Logotype design at the first generation Logotype design at the final generation	-0.9 2.2	$p<0.01$
7	Logotype design at the first generation Logotype design at the final generation	-1.0 1.8	$p<0.01$

Table 5. Comparison results of both experiments

Item No.		Experiments Types	Average Value	Significant Diffrenece
1	Layout	Experiment A Experiment B	1.9 0.2	$p<0.05$
3	Whole	Experiment A Experiment B	1.3 -1.3	$p<0.01$
	Shape of Letter	Experiment A Experiment B	1.9 -1.1	$p<0.01$
	Layout	Experiment A Experiment B	0.4 -1.4	$p<0.05$

5 Conclusions

This paper proposes a logotype design support system based on sketches. The proposed system generates design candidates using a sketch of desired logotype design and repeats design procedures using IGA. In the interaction with the system a user can modify design candidates using not only evaluation values of design candidates but also modification by sketch as necessary. Repeating the design procedures, a user can get desired logotype design. Subjects experiments are performed in order to verify usefulness of the proposed system. Experimental results show that the proposed system can design user's desired logotypes and that design operation by rough sketch is useful. Through the proposed system a user takes part in design actively by inputting rough sketches of logotype layouts and by modifying logotype design candidates using sketches. The followings are future problems: (1) Diversity of sketch input/modification is considered. (2) Color scheme of logotype design is introduced because color information of logotype can give many image to the people. (3) Subjects with experiences of logotype design perform the experiments and their comments are collected.

References

1. Yamamoto, S., Onisawa, T.: Interactive Support System for Logotype Design Based on User's Feelings. In: Proceedings of the Human Interface Symposium 2009, pp. 697–704 (2009)
2. Maeda, Y., Miyashita, S.: Chaotic Interactive Sound Generation System Using Interactive Genetic Algorithm. Journal of Japan Society for Fuzzy Theory and Intelligent Informatics 21(5), 768–781 (2009)
3. Cleveland, P.: Style Based Automated Graphics Layouts. Design Studies 31(1), 3–25 (2010)
4. Lee, J., Chang, M.: Stimulating Designers' Creativity Based on a Creative Evolutionary System and Collective Intelligence in Product Design. International Journal of Industrial Ergonomics 40, 295–305 (2010)
5. Hsiao, S., Hsu, C., Tang, K.: A Consultation and Simulation System for Product Color Planning Based in Interactive Genetic Algorithms. Color Research & Application (2012), doi:10.1002/Col. 21730
6. Hotta, H., Nozawa, T., Hagiwara, M.: An Automatic Japanese Font Designing System Using Kansei Rule Base. Journal of Information Processing of Japan 48(3), 1491–1501 (2007)
7. Kuroda, H., Ozawa, T., Kameda, H.: Implementation and Evaluation of a Web Page Design System Based on Impression Words. Technical Report of The Institute of Electronics, Information and Communication Engineers 107(387), 2059–2062 (2007)
8. Sakurai, Y., Tsuruta, S.: A Searching Method Based on Mechanically Learned Subjective Image. Journal of Japan Society for Fuzzy Theory and Intelligent Informatics 21(2), 214–221 (2009)
9. Morita, R., Onisawa, T.: Phrase Animation Generation Using Impression of Adjective/Adjective Verb. Journal of Japan Society for Fuzzy Theory and Intelligent Informatics 22(1), 121–134 (2010)

10. Tatara, T., Ohsaki, G.: Use of Global and Local Features in Sketch Image Retrieval System. The Institute of Image Information and Television Engineers 62(12), 2059–2062 (2008)
11. Sousa, P., Fonseca, M.J.: Sketch-based Retrieval of Drawings Using Spatial Proximity. Journal of Visual Language & Computing 21(2), 69–80 (2010)
12. Igarashi, T.: Recent Trends in Sketching Interface. Computer Software 16(5), 78–83 (1999)
13. Shesh, A., Chen, B.: SMARTPAPER: An Interactive and Use Friendly Sketching System. Computer Graphics Forum 23(3), 301–310 (2004)
14. Olsen, L., Samavati, F.F., Sousa, M.C., Jorge, J.A.: Sketch-based Modeling: A Survey. Computer & Graphics 33(1), 3–13 (2006)
15. Cluzel, F., Yannou, B., Dihlmann, M.: Using Evolutionary Design to Interactively Sketch Car Silhouettes and Simulate Designer's Creativity. Engineering Applications of Artificial Intelligence 25(7), 1413–1424 (2012)
16. Suwa, M.: Constructive Perception: Coordinating Perception and Conception Toward Acts of Problem-finding in a Creative Experience. Japanese Psychological Research 45(4), 221–234 (2004)
17. Obscure Inc. Idea of Logo Design! Learn from Example !! Professional Design Rule & Technique. Stock Company Kosaido (2010)
18. Yoshida, K., Nakagawa, Y., Koppen, M.: Interactive Genetic Algorithm for Font Generation System. In: Proc. of World Automation Congress(WAC) 2010, IFMIP538, pp. 1–6 (2010)
19. Kitano, H.: Genetic Algorithm 4. Sangyo Tosho (2003)
20. Unehara, M., Onisawa, T.: Music Composition System with Human Evaluation as Human Centered System. Soft Computing 7, 167–178 (2003)
21. http://www.logo-kako.com/

An Artificial Neural Network Based on the Architecture of the Cerebellum for Behavior Learning

Kenji Iwadate, Ikuo Suzuki, Michiko Watanabe, Masahito Yamamoto, and Masashi Furukawa

Dept. of Computer Science, Kitami Institute of Technology
165 Kouen-cho, Kitami-Shi, Hokkaido 090-8507, Japan
iwadate@mail.kitami-it.ac.jp

Abstract. In the last decade, artificial intelligence (AI) pervades every aspect of our lives. However, there is a gap between AI-based machine behavior and human in natural communication. The behavior of most AI is determined as a task list generated by engineers, but to obtain high-level intelligence, AI needs the ability to cluster tasks from circumstances and learn a strategy for achieving each task. In this study, we focus on the human brain architecture that gives it the ability to self-organize and generalize sensory information. We propose an Artificial Neural Network (ANN) model based on that architecture. We describe a cerebellum-based ANN model (C-ANN) and verify its capacity to learn from the phototaxic behavior acquisition of a simple two-wheeled robot. As a result, the controller of the robot is self-organized to be simple and able to achieve positive phototaxis. This result suggests that the proposed C-ANN model has the capability of supervised learning.

Keywords: artificial neural network, spike timing dependent plasticity, cerebellum, self-organization, behavior learning.

1 Introduction

Half a century has passed since a field of artificial intelligence (AI) had explored. In the last decade, AI has come to pervade every aspect of our lives. For instance, AI succeeded in beating human in a kind of game such as Chess (IBM, DeepBlue, 1997), Japanese chess (Information Processing Society of Japan, Akara 2010, 2010), and also quizzes (IBM, Watson, 2011). On the other hand, as use of smart-phones became widespread, the natural language processing system with voice known as Siri (apple, 2012). However, a gap still exists between AI-based machine behavior and human in natural communication. One of the major reasons is that a behavior of the most conventional AI-based machine is determined in advance as a task list generated by engineers. Although the AI-based machine is able to select a predetermined task well, it is unable to adapt a new task in a new circumstance.

Y.I. Cho et al. (eds.), *Soft Computing in Artificial Intelligence*,
Advances in Intelligent Systems and Computing 270,
DOI: 10.1007/978-3-319-05515-2_13, © Springer International Publishing Switzerland 2014

To obtain high-level intelligence, the AI-based machine should have an ability to cluster tasks from circumstances and to learn a strategy for achieving each task.

In this study, we focus on the architecture of the human brain that gives it the ability to self-organize and generalize sensory information. We propose a Neural Network (NN) model based on the human brain. In this paper we describe a basic NN model based on the structure and learning rules of the cerebellum (C-ANN) and verify its ability to learn.

2 Architecture of the Cerebellum

2.1 Network Structure

The cerebellum has an ancient origin in brain history, and it dominates action critical for life, such as understanding the current situation and learning locomotive behavior. It is too soon to know whether there exists a neuron in the cerebellum called the Purkinje cell that is involved in error information. Scientists expect that the network topology of the cerebellum can perform supervised learning by using signals from Purkinje cells as a training signal. In the 1980's, Masao Ito clarified the structure of the cerebellum, and the resulting theory of cerebellar function was widely accepted. Figure 1 shows the abstract structure of the cerebellum.

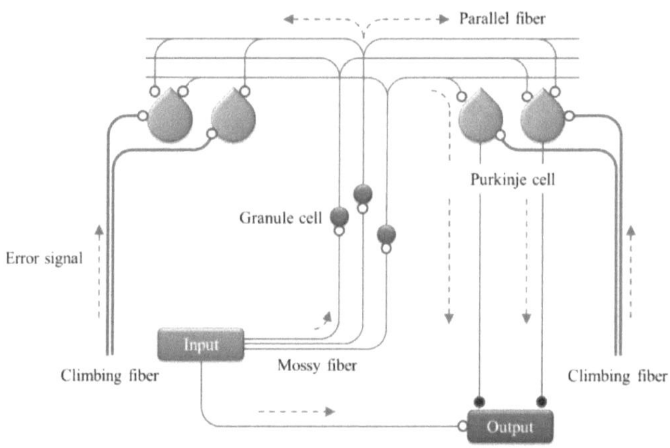

Fig. 1. Structure of the cerebellum

Two kinds of neurons called granular cells (GC) and Purkinje cells (PC) are involved in the learning process of the cerebellum. GC work as an input of the cerebellum. GC receive signals from sensors and are linked to PC with excitatory synapses. PC generate the output of the cerebellum. PC receive signals in two ways; one is input signals propagated from GC, and the other is an error signal from a climbing fiber (CF). PC have only inhibitory synapses that inhibit the output signal when they receive an error signal. At the same time, PC also damp signals from GC associated with the error signal. Despite the cerebellar theory described above is well known in biological science, there is little information about simulation or the learning ability of a cerebellar model. We aim to verify the learning capability of the cerebellar model so-named C-ANN.

2.2 Spike Timing Dependent Plasticity (STDP)

Spike Timing Dependent Plasticity (STDP) is a learning rule of the synaptic weight between neurons, which is proposed by Bi and Poo in 1998 [1]. STDP includes a traditional synaptic weight-learning rule that is well known as "Hebb's rule". Figure 2 shows the relationship between the time subtraction of neuron spikes (Δt) and the variation of the synaptic weight of a neuron-A that is connected to a neuron-B (Δw). Δw increases when the time subtraction has a positive value and decreases when the time subtraction has a negative value. In addition, Δw increases when the time subtraction is equal to zero, and it conforms to Hebb's rule.

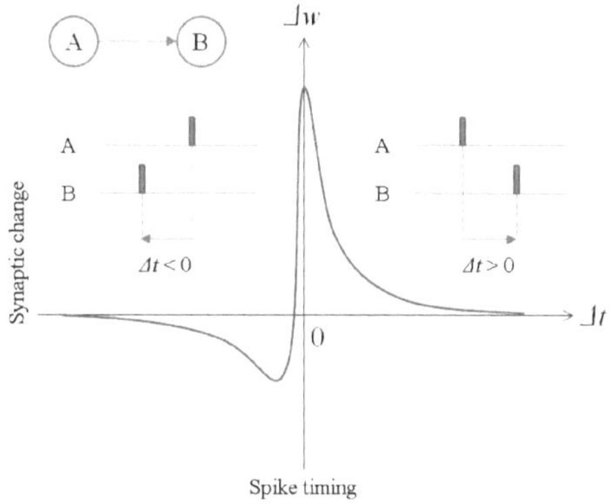

Fig. 2. Spike Timing Dependent Plasticity

3 Neural Network Model Based on Cerebellum Architecture

A cerebellum artificial neural network model (C-ANN) is discussed in this section. A model of a single neuron, a structure of the cerebellum, and STDP are described.

3.1 Model of a Single Neuron

A single neuron is modeled as a time series function based on the integrate-and-fire model for adopting STDP for learning [2]. Figure 3 shows the single-neuron model. The neuron has a number of synapses connected to other neurons as illustrated in Fig.3 (a). Each synapse has the different strength w_{ij} independently.

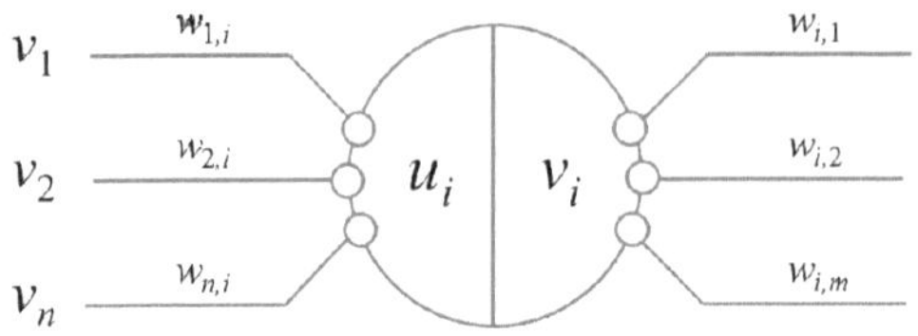

(a) Inputs and outputs

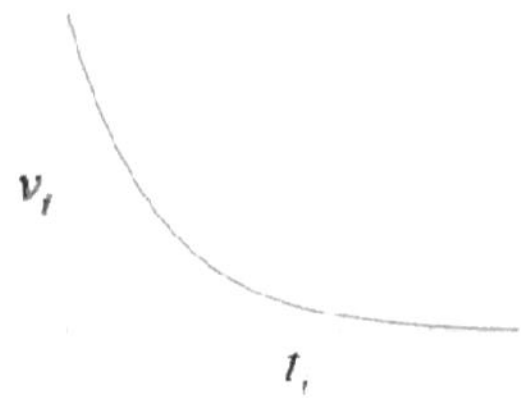

(b) A function of the output

Fig. 3. A model of the single neuron

The input of the neuron u_i is represented by Eq. (1) where Γ_i is the class of neighbor neurons of neuron-*i*, w_{ji} is the strength of the synapse from neuron-*j* to neuron-*i*, and θ_i is a spike threshold. The output of neuron v_i is formulated as a time-decayed function, as in Fig.3(b) and Eqs. (2)-(3), where t_i is elapsed time from a neuron spike, and τ_{vi} is the damping coefficient of the exponential function. t_i is reset to zero with a probability $p(u_i)$ shown in Eq.(3), where τ_i is a temperature coefficient in the sigmoid function Eq. (4).

$$u_i = \sum_{j \in \Gamma_i} (w_{ji} v_j) - \theta_i \tag{1}$$

$$v_i = \exp(-t_i / \tau_{vi}) \tag{2}$$

$$t_i = \begin{cases} 0 & (\text{w.p. } p(u_i)) \\ t_i + 1 & (\text{otherwise }) \end{cases} \tag{3}$$

$$p(u_i) = \{\exp(-u_i / \tau_i) + 1\}^{-1} \tag{4}$$

3.2 The Network Model of the Cerebellum

The C-ANN model is based on the structure of the cerebellum described in chapter 2. Figure 4 shows the designed structure of the C-ANN. The model consists of an input layer, a granular cell layer, a Purkinje cell layer, and an output layer. The strength of each synapse is defined as $w_{ij} \in [-1,1]$, and these values are initialized to zero. Neurons of a granular cell layer receive signals from the input layer and generate excitatory spikes to a Purkinje layer. Purkinje cells receive excitatory signals from the granular cells and an excitatory error signal from the climbing fiber. Purkinje cells generate inhibitory spikes to neurons of an output layer. Neurons of an output layer receive excitatory signals from the input layer directly and inhibitory signals from Purkinje cells.

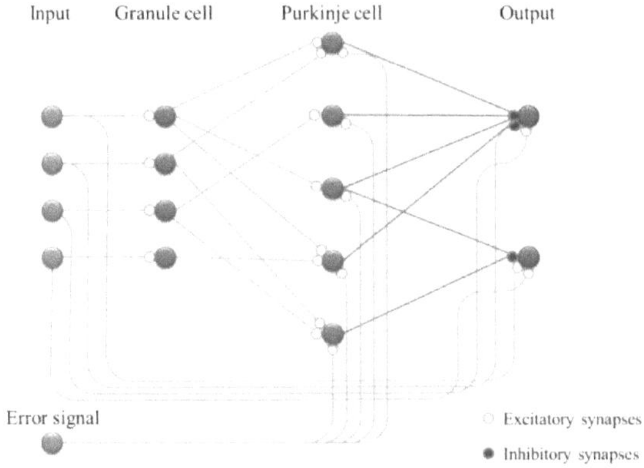

Fig. 4. A model of C-ANN

3.3 The Model of STDP

The amount of strength change of each synaptic connection Δw_{ij} is formulated as Eq. (5), where *A* is the gain of the change.

$$\Delta w = \begin{cases} Av_i & (\Delta t_i \geq 0) \\ -Av_i & (\Delta t_i < 0) \end{cases} \tag{5}$$

Figure 5 illustrates original and modeled STDP. A chained line shows original STDP, and solid lines show the modeled version.

4 Behavior Acquisition of the Virtual Robot

To verify the learning capacity of the proposed C-ANN, behavior acquisition of a virtual robot with the C-ANN as the controller is conducted. Acquisition of phototaxic behavior is specified as the task for the robot.

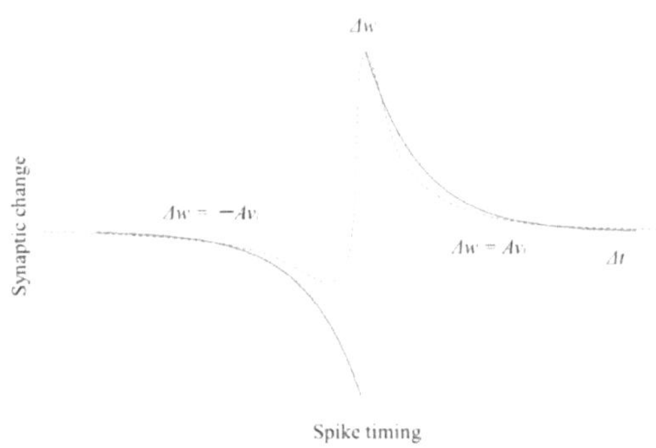

Fig. 5. A model of STDP

4.1 The Virtual Robot

A two-wheeled robot shown in Fig. 6 is adopted as the virtual robot. The robot has two types of optical sensor arrays and two motors for driving wheels. Each sensor array consists of 32 partial optical sensors. A front array detects light strength as an input signal. A rear array detects light strength as an error signal. Assuming that the proposed C-ANN has a structure to avoid the error signal, the robot comes to acquire a positive phototaxis.

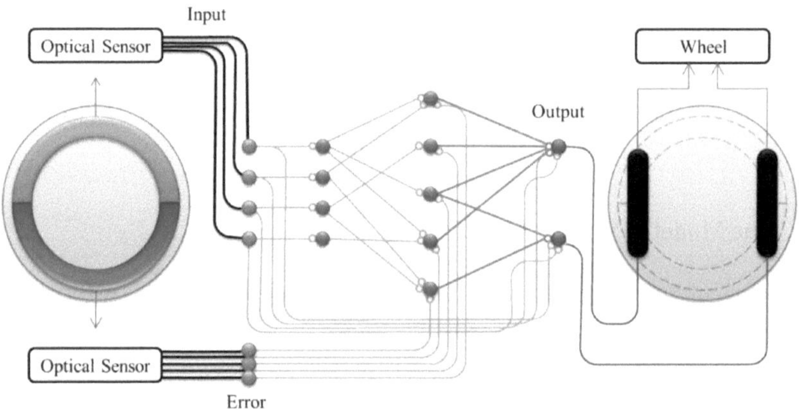

Fig. 6. The virtual robot

An input signal of each sensor (L_i) is calculated according to Eq. (6), where θ_i is the angle between the sensor and the light source, and N_s is the number of the sensor. Two wheels are connected to the output layer of the C-ANN, and these are driven according to Eq. (7), where v_t is the output of the C-ANN, and α is its inertia coefficient.

$$L_i = \begin{cases} \cos(N_s\theta_i) & -\pi/2 \leq N_s\theta_i \leq \pi/2 \\ 0 & \text{otherwise} \end{cases} \tag{6}$$

$$\omega_t = \alpha v_{t-1} + (1-\alpha)v_t \tag{7}$$

4.2 The Environmental Condition for Behavior Acquisition

Figure 7 shows the environment for behavior acquisition. The environment is represented as a two-dimensional field. The virtual robot is put at the origin of the field. In addition, a light source is put on the circumference of a circle in the environment. The position of the light source is changed every 5000 steps (1 trial). The robot evolves its behavior by continuing to change the strength of each synaptic connection of the C-ANN using only using STDP in every simulation step. Parameters of the robot and the C-ANN for simulation are set as Table 1.

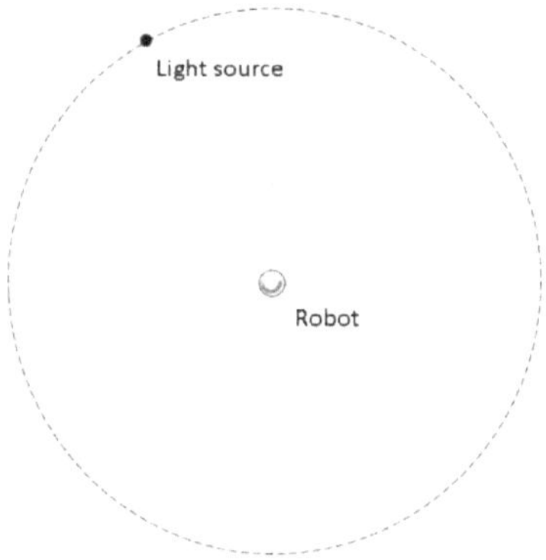

Fig. 7. Virtual environment

Table 1. Parameters of the robot and the C-ANN

Parameters of the robot	Parameters of the C-ANN
Diameter: 100 [mm]	The number of the input: 32
Distance between wheels: 40 [mm]	The number of the granular cell: 32
Max velocity: 100 [mm/sec]	The number of Purkinje cell: 32
Time resolution: 1/100 [-]	The number of the output: 2

4.3 Experimental Result and Discussion

Figure 8 illustrates trajectories of a robot as acquired behaviors and adapted C-ANN controller of the robot in the initial trial, the middle trial, and the final trial of the simulation. Filled circles show light sources put in the field, and solid lines show trajectories of robots. In the initial trial, the cumulated moving distance of the robot is low, because the strength of every synaptic connection is initialized to zero. The acquired behavior becomes more sophisticated as the simulation progresses. In the final trial, the robot attains a positive phototaxis completely and circles around the light source after it reaches to the light source. Moreover, the specified strength of synaptic connections for acquiring a positive phototaxis, shown as chained lines in Fig. 8(c) (i.e., connections between left-side sensors and the right motor and between right-side sensors and the left motor), increases. As a result, since synapse connections are self-organized only by STDP, a smart controller is obtained.

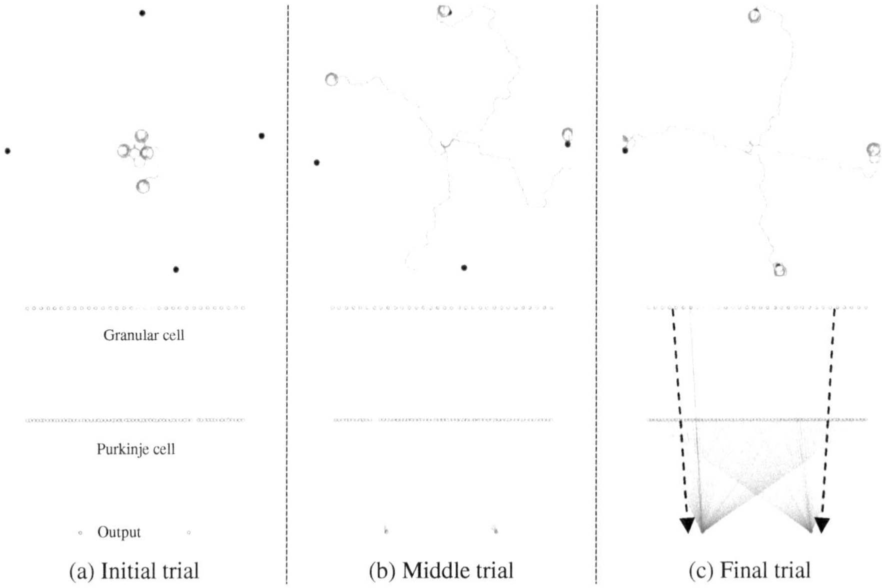

Fig. 8. Acquired behavior and controller

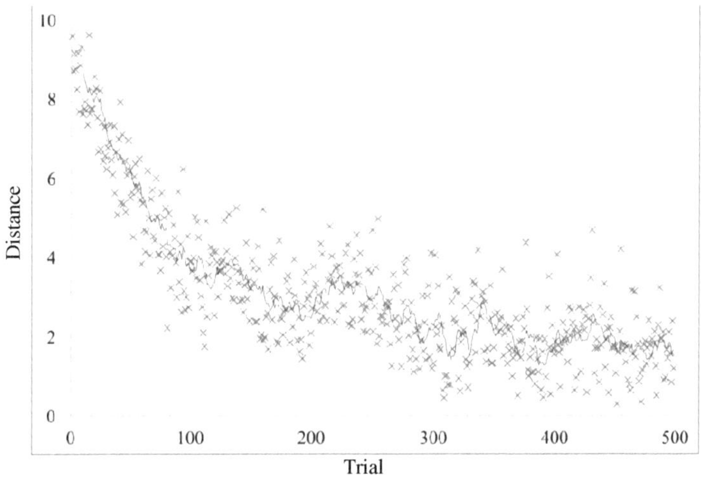

Fig. 9. Distance between the robot and the light source

Figure 9 shows the average distance between the robot and the light source in 10 simulations. The distance decreases during the simulation process. These results confirm the proposed model's learning capability.

5 Conclusion

In summary, we proposed the artificial neural network based on the architecture of the cerebellum, and we verified its ability to learn behavior via the behavior acquisition of a two-wheeled robot.

This result suggests that the proposed C-ANN model is capable of supervised learning using a signal from Purkinje cells as the error signal.

In future work, detailed analysis of the mechanism of supervised learning, an investigation of the ability of incremental learning, and coordination with a cerebrum model for acquiring high-level intelligence will be considered.

References

1. Bi, G., Poo, M.: Synaptic modifications in cultured hippocampalneurons: dependence on spike timing, synaptic strength, and postsynaptic cell type. The Journal of Neuroscience 18(24), 10464–10472 (1998)
2. Ezequiel, A., Paolo, D.: Evolving spike-timing-dependent plasticity for single-trial learning in robots (published online August 18, 2003)
3. A theory of cerebellar cortex. Journal of Physiology 202, 437–470 (1969)
4. A theory of cerebellar function. Mathematical Bioscience 10, 25–61 (1971)
5. Yamazaki, T.: On Modern Cerebellar Perception Hypothesis. The Brain and Neural Networks 18(1), 22–30 (2011)
6. Cyr, A., Boukadoum, M.: Classical conditioning in different temporal constraints: and STDP learning rule for robots controlled by spiking neural networks. Adoptive Behavior 20(4), 257–272 (2012)
7. Ferrari, S., Mehta, B., Muro, G.D., Van Dongen, A.M.J., Henriquez, C.: Biologically Realizable Reward-Modulated Hebbian Training for Spiking Neural Networks. International Joint Conferens on Neural Network, 1780–1786 (2008)

Author Index

MIX
Papier aus verantwortungsvollen Quellen
Paper from responsible sources
FSC® C105338
FSC
www.fsc.org

If you have any concerns about our products,
you can contact us on
ProductSafety@springernature.com

In case Publisher is established outside the EU,
the EU authorized representative is:
Springer Nature Customer Service Center GmbH
Europaplatz 3, 69115 Heidelberg, Germany

Printed by Libri Plureos GmbH
in Hamburg, Germany